Air Pollution

Environmental Issues Series
Scientists' Institute for Public Information

Edited by Barry Commoner

Air Pollution

Virginia Brodine
Consulting Editor,
Environment

Harcourt Brace Jovanovich, Inc. HBJ
New York Chicago San Francisco Atlanta

Cover photo by Harvey Stein
Photo credits and copyright acknowledgments on page 193

ISBN: 0-15-502112-5

Library of Congress Catalog Card Number: 72-94156

Printed in the United States of America

OSU?E MODEL VALIDATION STUDY AT THE
LEAR GENERATING PLANT Washington : U.S.
l Protection Agency, Office of Radiation
76. iv, 27 p. : ill. ; 27 cm. (EPA ; 520/

es.
8
: p. 25-27.
--Physiological effect. 2. Radiation--Safety
-Minnesota. I. Partridge, J. E. II. United
ronmental Protection Agency. Office of
ograms. III. Monticello Nuclear Generating
aul, Minnesota.
31350216

6.2:N 88
Point Gov. Doc. EP 6.2:N 88

BREWSTER FLIES FOR THE U. S. COAST GUARD,
Bolles, Air patrol; Jim Brewster flies for
ast guard, by Henry B. Lent, illustrated with
S. Coast guard photographs. New York, The
mpany, 1942. 170 p. 1 l., incl. front.,

ates. Coast guard--Juvenile literature.
42-22864 25285653
L +
PL + 614 L
S 359.9 LE BOOKS
J 629.13 LE

A 501 PIPER PA-31-350, N5MS, PHILADELPHIA,
JULY 25, 1980.
s. National Transportation Safety Board.
ident report : Air Pennsylvania 501 Piper PA-
, Philadelphia, Pennsylvania, July 25, 1980 /
nsportation Safety Board. Washington, D.C.
; [Springfield, Va. : National Technical
Service, distributor], 1981. ii, 24 p. :

-1."
cs--Pennsylvania--Philadelphia--Accidents.
ic turbulence. 3. Aeronautics--Pennsylvania-
a--Safety measures. I. Title: Air
501 Piper PA-31-350, N5MS, Philadelphia,
, July 25, 1980.
32072058
1.112:81-1

TY OF DEPOSITED SNOW.
omu, Air permeability of deposited snow.
pan, Institute of Low Temperature Science,
. illus. 26 cm. (Contributions from the
Low Temperature Science. Series A, no. 22)

TD
883
.B76

AIR POLLUTION.
Brodine, Virginia. Air pollution. New York, Harcourt
Brace Jovanovich [1973] xvi, 205 p. illus. 24 cm.
(Environmental issues series)
Includes bibliographical references.
1. Air--Pollution
m00668191 72-94156 20881916
Ref and Loan TD883 .B76
AR Beloit Col. Science 628.53 B784a
AR Beloit PL 614.71 B784
ES Sheboygan PL 614.71 B78 NONFICT
ES UWC-Sheboygan TD883 .B76
IN UW-Eau Claire TD883 .B76
IN UW-River Falls TD883 .B76
MC Manitowoc PL 628.5 B
MD Fond PL 628.53 B78
MD UWC-Fond TD883 .B76
MW Milw PL 628 53-B864 00788994
MW UW-Milwaukee TD883 .B76
NI Southern Door HS 628.5 BR
NI St Norbert Col 628.53 B78a MAIN
NI UW-GB TD883 .B76
NW UW-Superior TD883 .B76
SW UW-Platteville TD883 .B76
TC UW-Parkside TD883 .B76
WK Elm Grove PL 628.53 B864
WR Tomah HS 628.5 BRO
WR UW-La Crosse TD883 .B76
WR Western Wis Tech TD 883 B76 MAIN
WV Marathon Co. 614.7 B784a
WV Nicolet Col TD883 B76
WV UW-Stevens Point TD883 .B76

AIR POLLUTION
Bryson, Reid A. Air pollution [by] Reid A. Bryson and
John E. Kutzbach. Washington [Association of American
Geographers, 1968] 42 p. illus. 28 cm. (Association
of American Geographers. Commission on College Geography.
Resource paper no. 2)
Bibliography: p. 37-42.
1. Air--Pollution--United States. I. Kutzbach, John E.,
joint author. II. Series.
m00451528 68-54859 09456821
AR UWC-Rock Co. TD883.2 .B7
IN UW-Eau Claire TD883.2 .B7
MW UW-Milwaukee TD883.2 .B7
TC Carthage Col. TD 883.2 B916 A3
TC UW-Whitewater TD883.2 .B7
WR UW-La Crosse TD883.2 .B7

AIR POLLUTION [COMPUTER FILE]
Chandler, James R. Air pollution [computer file] / by
James R. Chandler. Pelham, N.Y. : Educational Materials
and Equipment, 1982. 1 program file (Apple) on 1

AIR POLLUTION ABATEMENT
National strategies
abatement : results
within the framework
Transboundary Air Po
1987. 56 p. : ill
"Sales No. E.87.II.
1. Air--Pollution--
pollution. I. Unite
Europe. II. Title:
m17065322 88-14
NI UW-GB Gov Pub Int

AIR POLLUTION ABATEMENT
AN INPUT-OUTPUT ANALYS
Miernyk, William H.
regional economic d
[by] William H. Mie
Mass., Lexington Bo
Bibliographies: p.
1. Air--Pollution--
Regional planning--
analysis I. Sears,
author.
m00898164 74-82
FV Lawrence U HC110.
FV UW-Oshkosh HC110.
IN UW-Eau Claire HC1
MW UW-Milwaukee HC11
NI UW-GB HC110.A4 M5
WV UW-Stevens Point

AIR POLLUTION ABSTRACT
1971-June 1976. R
U.S. Environmental
Quality Planning an
Supt. of Docs., U.S
Vols. for Feb. 1971
Pollution Technical
of Technical Inform
Technical Informati
Vols. for Feb.-Mar.
Protection Agency's
National Technical
Commerce; Apr. 1971
Technical Informati
Programs; Jan. 1973
Air Quality Plannin
Continues: NAPCA a
Key title: Air pol
Park. 1971) ISSN
I. United States. E
of Air Programs.
Protection Agency.

Preface

Environmental education at the college level is in a state of change, experimentation, and expansion. As a result, there is a great variety in course offerings and a desperate shortage of appropriate textbooks. *Air Pollution* differs from other books on the subject in purpose: The book marshals the current scientific knowledge that is relevant to the social issue of air pollution, showing how air pollution affects the earth-atmosphere system, how it affects the human body throughout its life span, and, finally, how these matters are related to the need for social and economic change. *Air Pollution* lays out the important facts, discusses the gaps in our knowledge, critically examines present efforts to solve the problem, and suggests the nature of the social decisions that lie ahead. No particular course of action is recommended. Instead, the student is confronted directly with a real and pressing problem and with the possibility that he or she may help to solve that problem. It is my hope that teachers who are concerned with making college education relevant to the daily lives of their students and to the development of a generation capable of dealing with the environmental crisis will find this book useful.

Many college teachers have written to *Environment* magazine expressing the need for a book of this kind. The magazine itself is widely used in environmental courses, and the aspects of its approach that have found favor with teachers and students have guided the organization and presentation of this material. Some of the scientists who worked on *Air Pollution* have developed environmental education programs in their own institutions, and their experiences have been helpful.

The book is designed to be understandable at the college level without previous science courses. Every effort has been made to provide sound, carefully documented scientific information and, at the same time, to avoid an academic, stilted style. This is exciting subject matter and there is no reason it should not be interesting reading. Students who have had chemistry, ecology, meteorology, or another of the many disciplines from which information has been drawn may find an occasional explanation unnecessary, but a less technical presentation makes the book usable in a wider variety of courses. Nor is the book limited to science curricula. It may be used by teachers and students concerned primarily with the sociology or the politics of pollution. As environmental issues come increasingly to the fore, the economic and political implications demand attention.

This book was made possible by the help of many scientists who advised, instructed, reviewed, criticized, and, in some cases, even wrote whole paragraphs.

The following members of the Air Pollution Committee, Scientific Division, St. Louis Committee for Environmental Information participated: Anna Coble, Department of Physics and Astronomy, Howard University; Robert Karsh, an internist in private practice; Robert E. Kohn, Assistant Professor of Economics, Southern Illinois University, Edwardsville; Albert J. Pallmann, Professor of Atmospheric Science and Chairman, Executive Committee, Institute of Environmental Studies, St. Louis University; John A. Pierce, Associate Professor of Internal Medicine and Chief, Pulmonary Disease Division, Washington University School of Medicine; Raymond G. Slavin, Associate Professor of Internal Medicine and Director, Allergy Research Laboratory, St. Louis University School of Medicine; and William M. Vaughan and Charles W. Lee, both with the Air Pollution Task Force of the Center for the Biology of Natural Systems, Washington University.

The following read all or part of the manuscript, made helpful comments, and identified specific problems: Bertram W. Carnow, Professor and Head, Occupational and Environmental Medicine, University of Illinois School of Public Health, and Professor of Preventive Medicine and Community Health at the same university's Lincoln School of Medicine; John R. Goldsmith, Head, Environmental Epidemiology Unit, California State Department of Public Health, and Vice President of the Northern California Committee for Environmental Information; Clarence C. Gordon, Director of Environmental Studies, University of Montana, and a member of the Western Montana Scientists' Committee for Public Information; Michael McClintock, Senior Scientist, Space Science and Engineering Center, University of Wisconsin, and a member of the Wisconsin Committee for Environmental Information; Glenn Paulson, Natural Resources Defense Council; Louis Slesin, a doctoral student at The Massachusetts Institute of Technology; and Lowell Wayne, Vice President and Director of Research, Pacific Environmental Services, Inc., Santa Monica, California. The participation of these scientists was invaluable; the responsibility for the final book and any errors therein remains with the author and editor.

To all my colleagues on the *Environment* staff, I am indebted for unflagging interest and frequent assistance. Kevin Shea answered technical questions, suggested sources, and reviewed drafts; Sheldon Novick, Julian McCaull, and Judith Meyer made helpful suggestions; Dianne Hoener and Tina Rehg assisted with picture research; Sherwood Novick located materials. Special thanks go to Diane Sullivan, who patiently and expertly transcribed very rough copy into tidy and accurate manuscript through repeated drafts.

Finally, it is my greatest pleasure to acknowledge the encouragement and support of my husband, Russell Brodine, throughout my work on this book, as in all my work.

Virginia Brodine

Foreword

This book could not have been published ten years ago. At that time air pollution was a subject that concerned only a small, specialized group of experts: environmental engineers, industrial physicians, and a few chemists. A book such as this one, which is intended for students generally, would have found no place in the curriculum that could accommodate it, much less demand it. Teaching, if not learning, was divided into neat, separate boxes, and a subject, if it was to be taught, had to fit into one of them.

In one dimension, this division was a division between disciplines—chemistry, biology, economics, history, and so forth. In another dimension, each of these disciplines, especially the sciences, has been further divided between "basic" and "applied" areas. For example, in biology the study of the exchange of oxygen between air and the blood in the lung is considered "basic," but the effect of smog on this process is an aspect of "applied" biology. The applied fields are usually regarded as intellectually less demanding than the basic ones, because a basic science defines its own problems (What is a gene? What force holds the atomic nucleus together?), whereas an applied science is obliged to solve problems that arise elsewhere (How can the yield of corn be improved? How can more powerful bombs be made?). Those who acknowledge that even basic science ought to have some social (as distinct from personal) purpose usually insist that useful applications will automatically arise out of basic studies, as unplanned fruits of the tree of knowledge.

A book about air pollution for the general student has little place in such a system of teaching. For the truth is that air pollution is a problem defined by society rather than by science. It is therefore, almost by definition, an area of "applied" rather than "basic" science.

And yet, in nearly all universities, colleges, and schools, the curriculum is still largely governed by the separate disciplines, and there remains a strong body of opinion that teaching air pollution to undergraduates is a regrettable departure from the intellectual rigor of a discipline—a temporary concession to a fad. What *has* changed is, first, that there is some willingness among their practitioners to relate the disciplines to one another. For the sake of dealing with an urgent problem such as air pollution, chemists, physicists, biologists, physicians, and engineers each seek out for study some appropriate part of the overall problem, which then becomes an *inter*disciplinary one. A second change has to do with the matter of relevance. In the last few years, students and many of their teachers have found the gap between the wealth and splendor of modern knowledge, especially in science, and the poverty and grimness of modern society too much to bear in silence. One outcome has been a growing demand for courses that make an effort to relate what we know to the good and evil that comes of knowing it—courses that deal explicitly with war, poverty, racism, and the environmental crisis.

While the movement toward interdisciplinary studies that are relevant to the urgent problems of society is encouraging, it leaves many questions unanswered. One of the virtues of Virginia Brodine's treatment of air pollution is that, apart from its analysis of the subject in its own right, it illuminates the more general issue of the relation between environmental problems such as air pollution and the basic disciplines. In its account of air pollution problems, the book reveals, it seems to me, strong reasons why the actual solution of the environmental crisis will require of the academic community much more than the creation of "interdisciplinary" or "relevant" courses or programs of research. What will be required of us, I believe, is not interdisciplinary studies but *adisciplinary* ones,* and not mere relevance to social issues but *participation* in them.

The interdisciplinary approach accepts the character of the separate disciplines—the methods and bodies of knowledge which they encompass—as given, and attempts to link them into a collaborative sum that relates to the various facts of a multifaceted problem such as air pollution. This approach assumes that in the discipline of chemistry, for example, as it is presently constituted, one can find the knowledge that can explain smog, and that some feature of biology, as it now exists, can explain how smog affects health. It suggests that while the disciplines must learn how to collaborate, they need not change internally in order to do so.

It seems to me that the evidence contradicts this assumption. This is not to say that environmental systems disobey the basic laws of physics, chemistry, or biology. Rather, the problem is that the laws of nature which we now know do not encompass all of nature and, in particular, may not, in their present form, be appropriate to the complex systems that operate in the environment.

Such a statement is often taken to mean that some form of mysticism (in biology, "vitalism") is being put forward. The concept is not that easily dismissed. The operational question is this: Is it *necessarily* true that all, or even the most important, properties of a complex system (i.e., one composed of interacting parts) can be discovered and understood by studying

* The important distinction between *inter*disciplinary and *a*disciplinary studies was suggested to me by Dr. Thayer Scudder of the California Institute of Technology, a gifted practitioner of the latter.

the properties of the separate parts? In more theoretical terms, the matter can be put this way: Given that studies of the material entities (atoms, molecules, organisms) found in nature, not only as separate units but also in complex associations (for example, an ecosystem), lead to the development of natural laws that account quite well for the properties of these separate entities, is it possible that new properties arise from the interactions among them and need to be accounted for by new natural laws? Or, put another way, the question is this: Given the laws of physics, chemistry, and biology devised from the study of atoms, molecules, and living organisms, can these laws successfully *predict* the properties of the complex systems (for example, urban air) in which all these things interact?

While this is not the place to argue the merits of these issues in general, it is appropriate to consider how they arise in environmental problems. There is, I believe, empirical evidence that the natural laws elucidated by the separate disciplines are insufficient to guide us toward solutions of most environmental problems. How else can we account for the fact that the environmental crisis has come upon us so suddenly? Clearly physics, chemistry, and biology have made enormous strides in the last 50 years. If the principles they have elucidated are indeed capable of predicting the appearance of urban smog and the carcinogenic effects of airborne asbestos particles, why were we taken by surprise by these and similar hazards? The fact is that the development of the environmental crisis was detected not by the application of the growing knowledge of atomic structure, molecular interactions, and biochemical processes, but by attention to the natural history of the environment as it has been changed by the hand of man. The environmental dangers of DDT were not detected by experts in the molecular structure or even the biological effects of such compounds, but by Rachel Carson and others concerned with the natural history of the environment. The carcinogenic hazards of asbestos were not predicted by the numerous studies of the chemical and physical properties of asbestos incident to its wide use in industrial technology, but by studies of the natural history of lung cancer.

Holism—the concept that because a complex system is characterized by the *interaction* among its parts, such a system cannot be described by summing up the properties of its *separate* parts—can of course be supported on purely philosophical grounds. But even on crudely empirical grounds, it would seem prudent to regard our present understanding of physics, chemistry, and biology as, at the least, an incomplete guide to the behavior of environmental systems. Hence, to return to the original point, if we now hope to develop a better scientific understanding of the hugely complex systems which are the arena of environmental pollution, some changes will have to occur *within* these separate disciplines.

The kinds of changes that need to take place are suggested by the environmental problems set forth in this book. In all of them, the keynote is complexity—systems of elaborately interacting parts that demand the kind of attention to the whole precluded by preoccupation with the parts. This, it seems to me, calls for science that is *adisciplinary* rather than *inter-*

disciplinary. By this I mean science based on the given problem that does not insist on breaking it down automatically into fragments which fit into the existing structures of the basic sciences. We know very little as yet about the practice of adisciplinary science. One of the virtues of this book is that it may encourage students, and perhaps some of their teachers, to take a hand in developing it.

Clearly this book should satisfy those who seek relevance in their academic work. But it suggests, as well, that relevance in our studies must be intimately connected with action. Among academics, the traditional approach to social action is that study comes first and action later. This attitude is solidly based in logic and in the traditional view that education is a *preparation* for life. Nevertheless, there are reasons to doubt that this approach will suffice to develop an effective understanding of environmental problems such as those discussed in this book.

Again, it is the intrinsic complexity of the environment that dictates a different approach. Consider the problem of improving the quality of urban air. First, there is the obvious link between the scientific analysis of air pollution and action to combat it: Clearly such action should be guided by the analysis. However, there is another, less obvious, link between studying and improving air quality. We know that urban air is a mixture of numerous interacting agents, so that any effort to reduce the concentration of any one of them (for example, waste hydrocarbon) is likely to alter the status of a number of other pollutants (for example, nitrogen oxides). Thus the purely scientific aspects of the problem—such as analysis of the composition of polluted air—will themselves be affected by almost any action taken to improve air quality. Clearly what is required is the most intimate interaction between efforts to study and efforts to improve the environment.

There is no simple formula for achieving such an interaction. One approach, which is the basis of the work of the Scientists' Institute for Public Information and its affiliated local groups, is for scientists who study the environment to pass that knowledge along to interested citizens and the community at large, so that their active efforts to improve the environment can be soundly based on knowledge. Another approach is to establish close links between the scientific studies of the environment—for example, university research projects—and official agencies working to improve the environment. Finally, there is the difficult but crucial effort to seek out, both in study and action, the links connecting the manifestations of environmental degradation in the natural world to the social processes—for example, the economic factors that tolerate, and even foster, greedy, short-term exploitation of resources—that are the root cause of the environmental crisis.

Here is the greatest challenge to the student of environmental degradation: To learn how to use this knowledge for its most important purpose, the advancement of human welfare.

Barry Commoner

Contents

Air Pollution

Episode 104

On August 17, 1969, hurricane Camille swept in from the Gulf of Mexico across the coastal area of Mississippi. Camille killed more than 200 people and destroyed almost $1\frac{1}{2}$ billion dollars worth of property—a devastating reminder that technological progress has not made man independent of nature.

A few days later, on Friday, August 22, the first signs of another kind of adverse weather condition were noted over parts of the Midwest. Skies were expected to remain sunny and temperatures, not unduly high for August, were not expected to rise. No storms were anticipated in the area, yet there were signals of adverse weather—reminders that *because of his technological progress* man is now at the mercy of the weather in a new way. In addition to forecasting sunny or cloudy skies and approaching storms, he must now keep a weather eye out for a new danger: High Air Pollution Potential (HAPP).

Such a potential was observed in the Great Lakes region that Friday, moving South on a broad front. This was to be called Episode 104 by federal officials, who warned that for a week or more following August 23, the sluggish dispersion of air pollutants could be expected throughout the area from the Mississippi Valley to the eastern coastal plains.

Human dependence on the weather to help rid cities of their daily burden becomes apparent when there is no rain or snow to wash pollutants out of the air, when there is no wind to blow them away, and when temperature and pressure hold pollution near the ground. HAPP is simply a weather pattern that intensifies everyday pollution.

Under normal conditions, temperature decreases with height for several miles above the earth's surface. If the ground is sufficiently heated by the sun, the air near the ground is warmed and pockets of air rise, bearing pollutants from traffic, industrial stacks, and other pollution sources. These upward-moving pockets of air are then replaced by cooler, cleaner air from above. In the presence of turbulent wind, this *vertical mixing* of air is increased. At night, the ground cools more rapidly than the air, consequently cooling the air near the earth's surface and reversing the usual daytime *vertical temperature distribution.* Within the first 1000 feet or so, the temperature may then *increase* with height, the so-called *temperature inversion* or *thermal inversion.* In this situation, instead of rising, the cool, pollutant-laden surface air is held near the ground by the warm air above it, but pollution builds up near its sources only if winds are absent or very light (that is, only if pollutants can move *neither* upward nor downwind). Nighttime surface inversions usually dissipate after the sun rises.

Inversions higher up may be much more widespread and longer lasting. For example, an area of high barometric pressure—air that is sinking and becoming warmer and denser—lies over the Pacific Ocean off the North American coast continuously during the warm months and frequently at other times of the year as well. This warm air does not sink all the way to the ground, but moves over Pacific coastal cities at altitudes of 2000–3000 feet, trapping air that has been cooled by ocean currents with its load of pollutants below. These high inversions often persist for days and are one of the reasons Los Angeles has such severe air pollution problems.

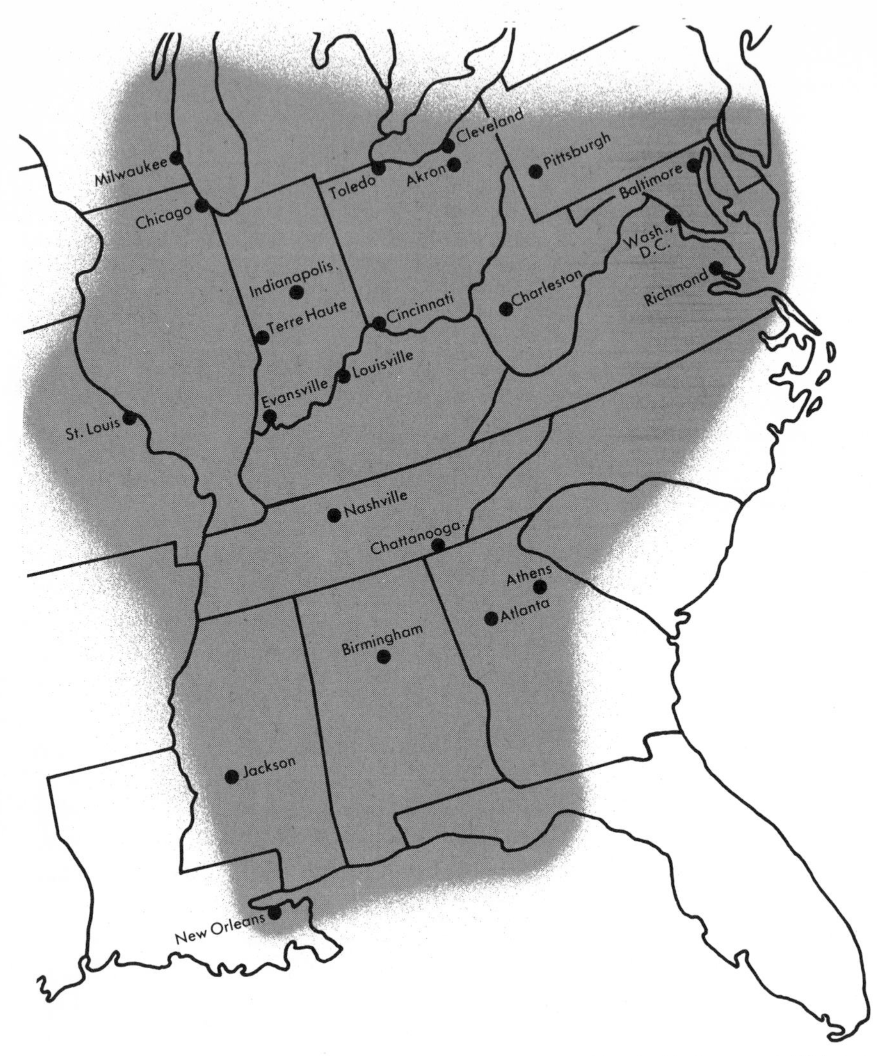

Episode 104 adapted from the Episode 104 map issued by the Environmental Science Services Administration.

Although the Midwest has no such semipermanent area of high pressure, migratory high-pressure systems called *anticyclones,* usually moving slowly eastward, are not uncommon there. Such an anticyclone was the cause of the high air pollution potential in August 1969.

High pressure systems are usually accompanied by periods of calm and very light winds, making conditions unfavorable for *both* upward and outward movement of pollutants. Moreover, hurricane Camille on the Gulf Coast was barely over when an intense low-pressure system *(cyclone)* in the tropical Atlantic developed into a northward-moving hurricane, followed by still another storm center off the coast of Georgia and South Carolina. These storms over the Atlantic formed an effective barrier to

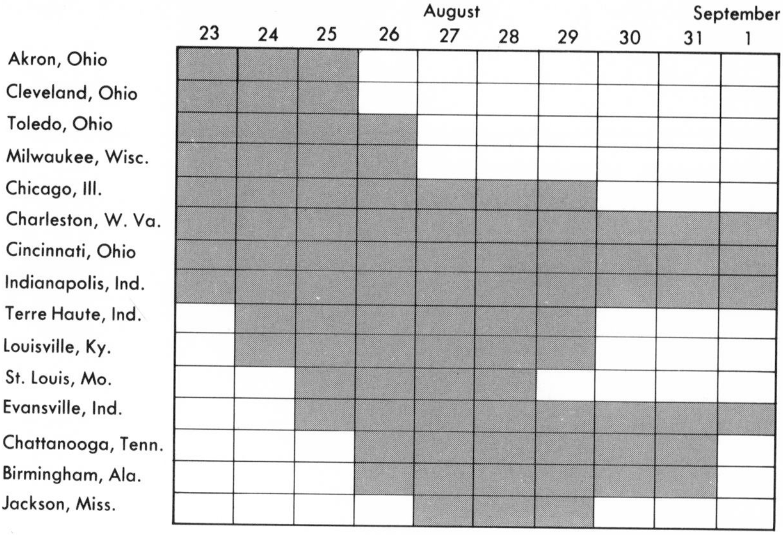

August September

23 24 25 26 27 28 29 30 31 1

Akron, Ohio
Cleveland, Ohio
Toledo, Ohio
Milwaukee, Wisc.
Chicago, Ill.
Charleston, W. Va.
Cincinnati, Ohio
Indianapolis, Ind.
Terre Haute, Ind.
Louisville, Ky.
St. Louis, Mo.
Evansville, Ind.
Chattanooga, Tenn.
Birmingham, Ala.
Jackson, Miss.

Cities for which data were available during Episode 104, showing days each was under the Environmental Science Services Administration's high air pollution potential (HAPP) advisory. In some of these cities (notably Milwaukee and Chicago), weather conditions unfavorable for the dispersion of air pollution began on Friday, August 22, before the first advisory was issued.

high-altitude air movement in a west-to-east direction. The Appalachian Mountains slowed the movement of the light winds toward the East and Southeast below 5000 feet. Most midwestern cities are built on low ground, often with hills or bluffs nearby which helped to keep pollution concentrated near its sources.

The bad air pollution weather which was to become Episode 104 was observed first in southern Wisconsin and northern Illinois on Friday, August 22. The Milwaukee County Air Pollution Control Department had just cited 58 industrial firms and government installations for violating the county standard for emission of *particulates*[1], the minute particles that give smoke its gray or black color. It was not surprising that haze and smoke began to cloud the city. On that Friday in Chicago, another typical city pollutant—sulfur dioxide—was more than triple what it had been the previous day and an air pollution alert was announced. A "watch" or first-stage alert requires no steps to reduce pollution, only a more careful monitoring of pollution levels so that any dangerous rise can be observed immediately.

By Saturday, the extent of the threatened area (shown in the preceding map and graph) had become clear. Indications were that the anticyclone would remain unbroken for several days. The National Meteorological Center issued its first HAPP advisory for Episode 104. Generally, south of Chicago, where the effects of the high-pressure system were just beginning to be felt, the weekend masked the problem—there were fewer pollutants to disperse—and activities went on as usual, with few people aware of the impending episode.

However, Monday morning, as the work week got under way and lines of traffic converged on the cities, a vast conglomeration of gaseous and particulate pollutants rose slowly and hovered in the still air under the inversion lid. St. Louis, like Chicago, was now under an air pollution alert.[2] Balloons carrying meteorological instruments, released at the St. Louis riverfront at 6:00 that morning, revealed that the city was lying under two inversion layers, shown in the following illustrations. There was a surface layer—rather typical of the early morning hours when skies are cloudless, humidity is low, and winds are light. There was also a higher layer of warm air associated with the high-pressure system spreading over the Midwest. By early afternoon, the surface layer had dissipated and there was a deeper *mixing layer*—a greater volume of air over the city was available in which the pollutants, now rising higher above their sources, could mix and become diluted. The dead calm of the early morning hours had given way to light winds from two to seven miles an hour, but the winds varied in direction from southeast to north: the southeastern wind blowing pollution from the industrial area across the Mississippi *toward* the city and the northern wind carrying pollution from power plants upriver *toward* the city center.

In spite of local differences in terrain, other cities in the area were having similar experiences with the weather and so with pollution. Weather bureau stations throughout the Midwest began to report reduced visibility due to smoke, haze, and, in the early morning, fog. The cities were hotter than the surrounding rural and suburban areas. (In St. Louis, minimum temperatures were from 3–6 °F higher at the riverfront, near the center of the city, than at the airport on its outskirts[3].) This has an effect on air circulation called the *heat island effect,* most noticeable at night and in the early morning—and at its maximum under highly polluted conditions like those of Episode 104. In such a situation, the polluted air may be trapped, forming a *dust dome* or *haze hood* which will not disperse because of the unfavorable weather. Instead a circulation pattern is set up which simply moves the polluted air up and outward to the edges of the dust dome; there some escapes, but much of the polluted air simply moves downward and back toward the center of the city. As light, variable winds carried the same air back and forth over pollution sources in cities affected by Episode 104, concentrations of pollutants built up[4].

With similar conditions prevailing throughout the Midwest, another effect of these intermittent light winds was probably to move pollution short distances outward from the industrial centers and traffic arteries and gradually spread it over the surrounding area. Indeed, as the stagnant high-pressure system persisted through the succeeding days, vacationers traveling homeward from the East or South toward Chicago and St. Louis observed a persistent haze in rural areas all the way across Kentucky and southern Indiana.

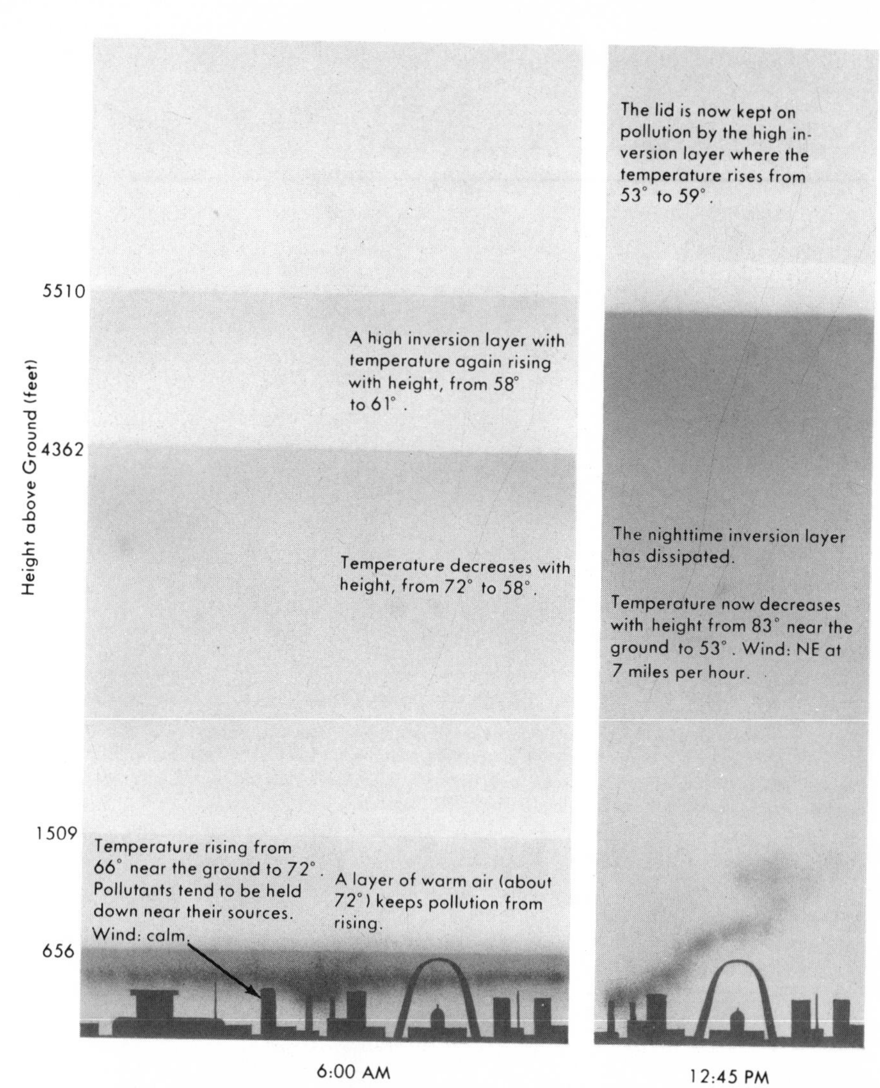

In the early morning hours, St. Louis and some of the other affected cities lay under two inversion layers. This drawing shows these conditions at the St. Louis riverfront on the morning and early afternoon of August 25, 1969.

No special technical knowledge was required to know that the air was now polluted. But just how polluted? That question proved difficult to answer. In Birmingham, Alabama, monitoring had been discontinued for lack of funds[5]. Evansville, Indiana, was in the process of moving its equipment and thus could do no monitoring. Jackson, Mississippi, with only a National Air Surveillance Network station for air sampling twice monthly, made only one measurement during the episode. Instrument trouble produced incorrect sulfur dioxide readings for most of the second half of August in Toledo. In Cincinnati, the technician in charge of the Continuous Air Monitoring

The air breathed by a child playing on a lawn in suburban Clayton, west of St. Louis, is very different from the air breathed by a child playing in Rush City, East St. Louis, in the shadow of the Sauget industrial section. In an episode, the pollution spreads to the suburbs, but is intensified in the industrial sections.

Program (CAMP) station was on vacation and the instruments measuring sulfur dioxide and nitrogen oxides were turned off (CAMP is a ten-station, federal network). Charleston, West Virginia, had discontinued particulate monitoring during a period of new efforts to reduce pollution (this monitoring was resumed in July 1970). Cleveland was monitored only twice a week. Some other cities in the area simply failed to respond to requests for information. The most complete information was available for St. Louis and for Chicago.

Even St. Louis, which has added additional monitoring stations since Episode 104, still has an incomplete picture of its air pollution, and this is hardly surprising. Consider what must be measured when the subject is *ambient* air—the air around us, a subject that has no beginning and no end, that is here today and gone tomorrow if the wind blows and here today and here tomorrow if the wind does not blow. But in the latter case, the *composition* of the air may be quite different tomorrow than today. Indeed, the air not only changes from day to day, but from one second to the next, and it may be vastly different in different parts of the same airshed or air quality control region.

On Thursday, August 28, when pollution concentrations were at their highest, the air breathed by a child playing on a lawn in suburban Clayton, west of St. Louis, was vastly different from that breathed by a child playing in a Rush City street in East St. Louis, a block or two from heavily traveled Illinois Route 3 and in the shadow of the Sauget industrial section which emits about 80 tons of sulfur dioxide into the air each day. Yet both children

How much dirtier the air became in Episode 104 is graphically illustrated in the comparison of episode particulates and average particulates in ten cities which appears on pages 10–11. Episode 104 concentrations are compared to nonepisode concentrations. The unshaded bars represent 1969 annual or six-month averages for each city where such averages were available. Where averages were not available, a comparable period just prior to the episode is used for comparison. The black bars show particulate concentrations on episode days, with the number indicating the date in August 1969. In some cities (Chicago, the St. Louis metropolitan area, Indianapolis, Cleveland, Akron, Louisville), there were reports from a number of different stations. In these cases, each bar represents the average of all stations in that city on the date indicated. Thus, the very highest particulate levels at single stations are not shown—for example, 593 micrograms per cubic meter on Sunday, August 24, at the Washington station in Chicago or levels above 400 on two successive days at the Carver station in the same city. Note also that in Chicago, particulates were monitored three times a week, not every day, even during the alert.

In Terre Haute, the air was sampled at some stations on one episode day and at others on another day. The black bars therefore represent a five-day episode average for seven city stations and for five rural stations. Episode averages are also shown where monitoring was done on more than one day but at only one station, as in Chattanooga. (Surprisingly, Chattanooga did not show an increase in particulates during the episode—the only city where this was the case.) In two cities (Cincinnati and Jackson) a report was available for only one day from one station.

To provide a bench mark for the evaluation of these measures of dirty air, the present federal standards are shown.

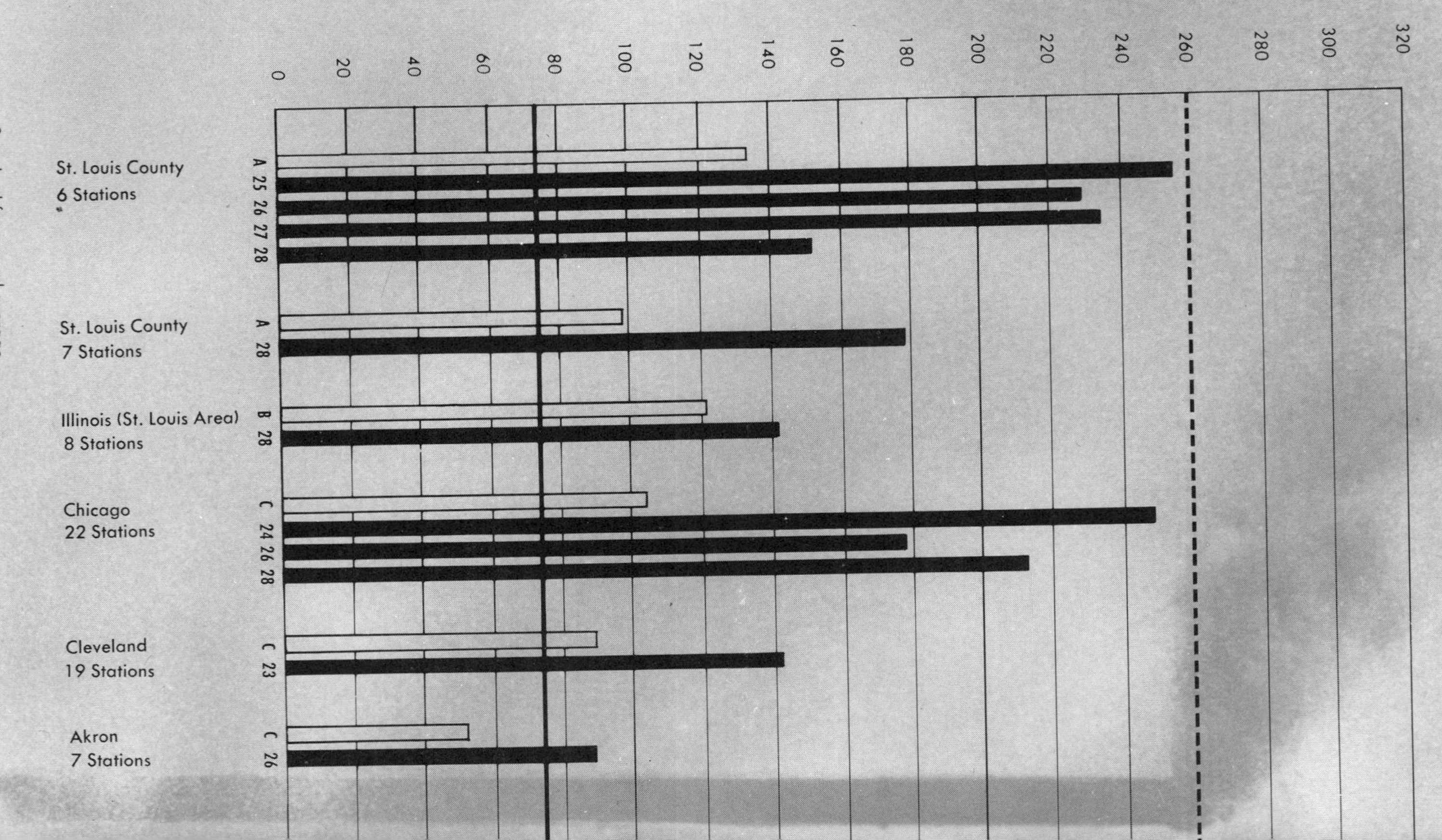

Standard for annual average

Standard, not to be exceeded more than once a year

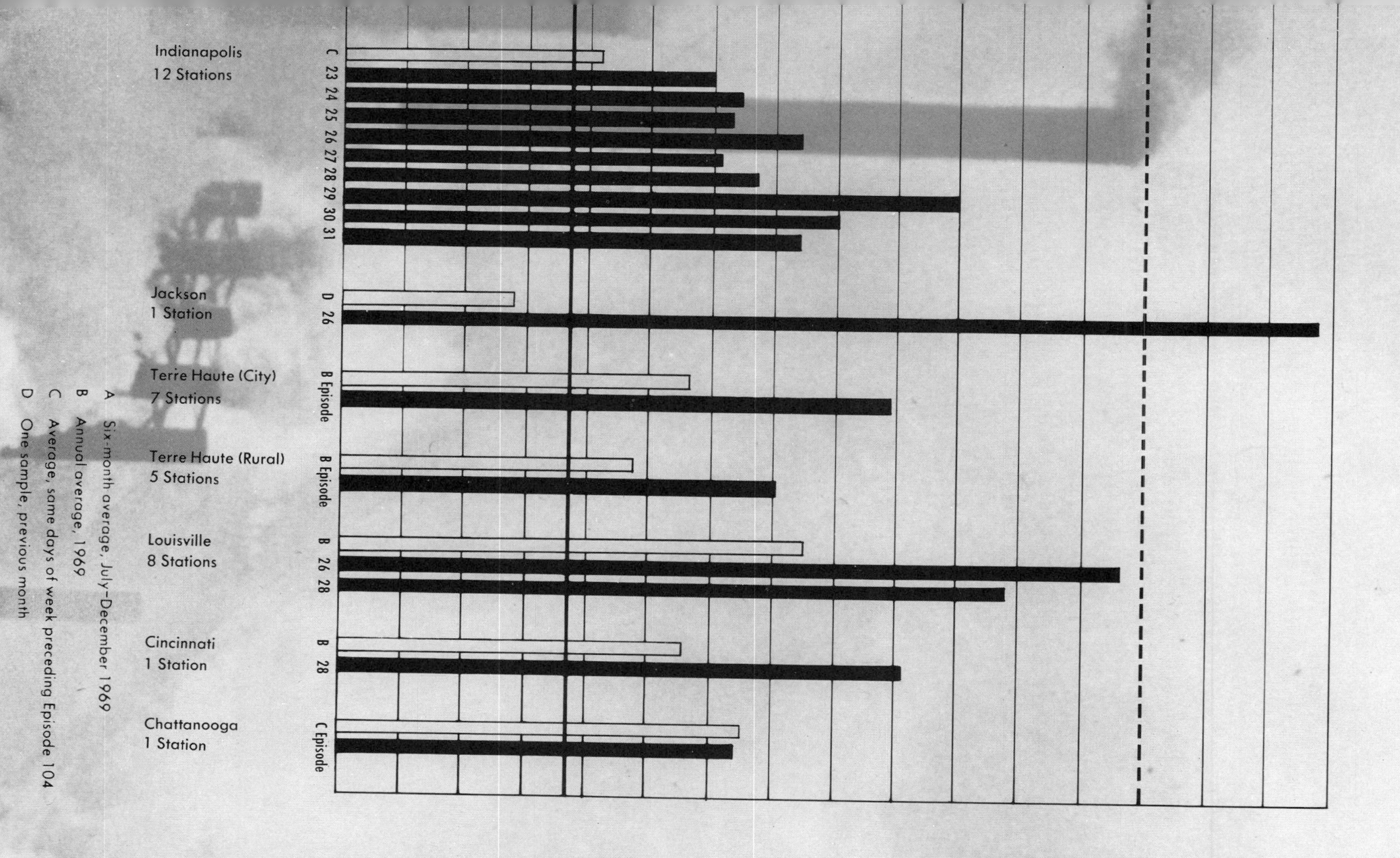

Indianapolis 12 Stations
C 23 24 25 26 27 28 29 30 31
Jackson 1 Station
D 26
Terre Haute (City) 7 Stations
B Episode
Terre Haute (Rural) 5 Stations
B Episode
Louisville 8 Stations
B 26 28
Cincinnati 1 Station
B 28
Chattanooga 1 Station
C Episode
A Six-month average, July–December 1969
B Annual average, 1969
C Average, same days of week preceding Episode 104
D One sample, previous month

were breathing the ambient air of the city. At any one place and at any one time in a highly industrialized urban center, a sample of this ambient air may contain a bewildering variety of both gaseous and particulate pollutants.

Increase in Pollutant Concentration

Particulates and Sulfur Pollution. Chicago remained under an air pollution alert until Friday, August 29. This city was rated second only to New York when the dubious honors of high pollution were handed out by the National Air Pollution Control Administration (NAPCA)[6]. When weather conditions were poor, pollution in Chicago began its rise from an already high level. Take particulates, for example—the tiny solid and liquid particles of smoke, metals, and other kinds of plain and fancy dirt that remain suspended in city air (larger particles of soil, flyash, and street dirt usually fall out rapidly and therefore are not included in measurements of suspended particulates). Air in a remote, nonurban area may have as little as 10 to 20 micrograms (millionths of a gram) of suspended particulates per cubic meter of air[7]. The federal standard* is now 75 micrograms, but in 1969 this was met during only one month at one Chicago station. In Episode 104, particulate concentrations reached a maximum of 593 micrograms per cubic meter of Chicago air.

Particulate concentrations rose throughout the area during Episode 104 (as shown in the preceding figure), but reported concentrations from one city are only roughly comparable to those from another. When particulate samples are subjected to laboratory analysis, scientists find considerable differences in the respective amounts of various kinds of particles because of the differences in industrial sources, and more than half the total composition of urban particulate samples remains unidentified[8]. Since certain kinds of particles are more dangerous to human health than others, this is not an academic matter. (The health effects of particulates are discussed later in "The Burdened Human.")

There appears to be not only a city-to-city difference, but also a difference between episode and nonepisode samples from the *same* city. A

* There were no federal standards at the time of Episode 104, but they have since been adopted for major pollutants. They are listed in the appendix on page 188.

How sulfur dioxide rose during Episode 104 in five cities. Episode 104 concentrations are compared to nonepisode concentrations. Data are from only one station in each city, except in Chicago and in Cleveland where values from eight and ten monitoring stations, respectively, are averaged. Unshaded bars represent six-month averages (July to December 1969 in the case of St. Louis), and averages for a comparable pre-episode period in the other cities. Again, federal standards are used as a bench mark. Sulfur dioxide levels generally rise and fall during each day, with the highest concentrations usually occurring near midday.

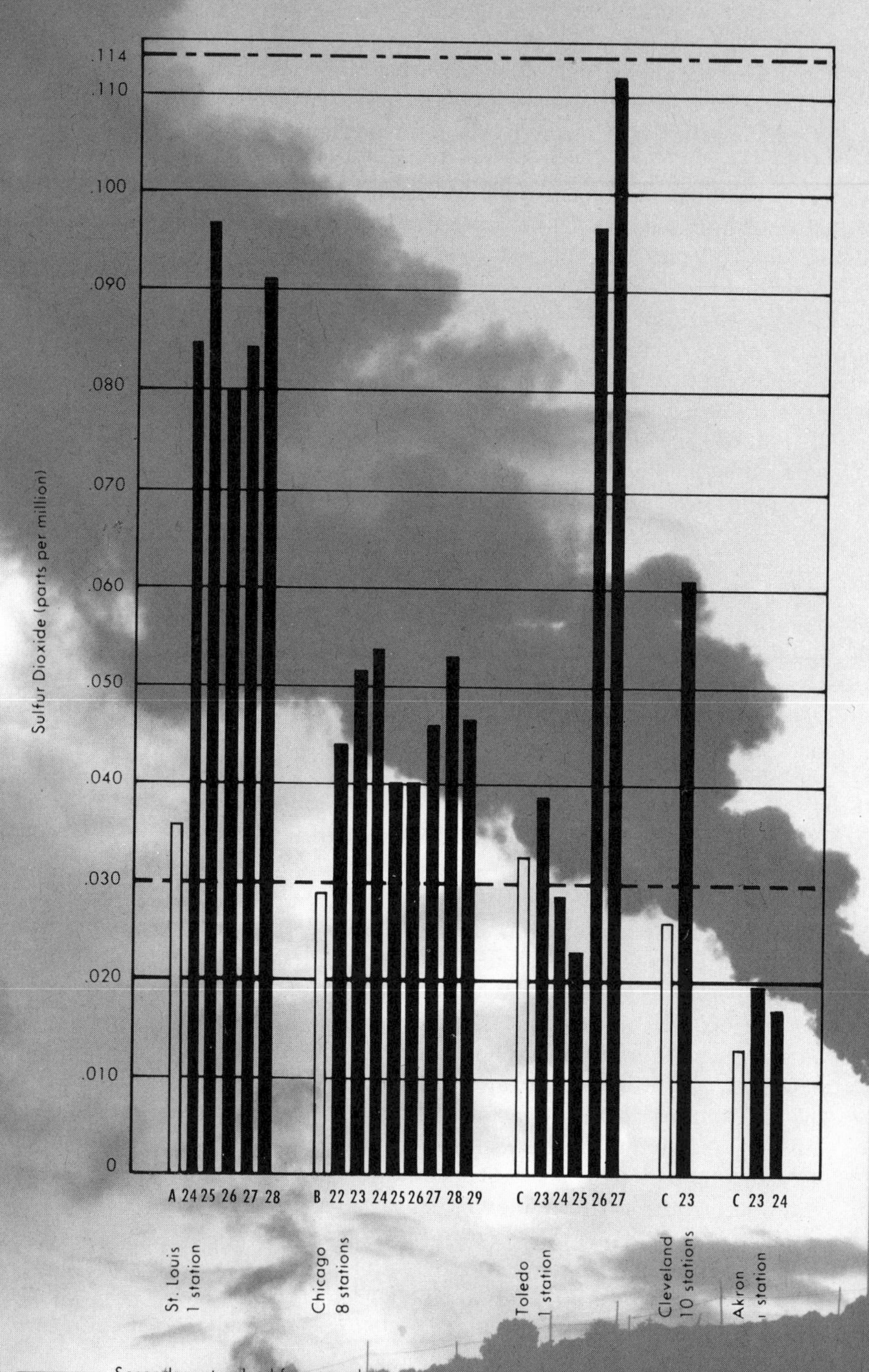

Secondary standard for annual average
Primary standard for annual average
Maximum 24-hour concentration, not to be exceeded more than once a year

A Six-month average, July-December 1969
B Average, August 22-29, 1966-1968
C Average, same days of week preceding Episode 104

Public Health Service pamphlet analyzing Episode 29 in November–December 1962, showed that not only did total particulate levels rise in affected cities, but more of the particulates were organic. In the New York–Boston area, the total number of particulates was three to four times higher than normal and the benzene-soluble organic particulates were seven times higher than normal[9]. This is cause for concern because *organic particulates* are biologically active and some may be carcinogenic.

It would be useful to know what happened to organic particles during Episode 104, especially since Chattanooga, St. Louis, Indianapolis, and Nashville have high concentrations of organic particulates under ordinary circumstances. All in the episode area, these four ranked among the top seven of 60 cities where samples were analyzed for organic particles in 1961–1965. However, a plan to analyze episode samples for organics, announced by the Public Health Service in 1964, was never carried through.

Sulfur dioxide data gathered during Episode 104 were available from five cities and also showed increases (see page 13). Particulates and sulfur dioxide are what might be called "old-fashioned" air pollution, typical of industrial towns and cities since the Industrial Revolution. Of course, quantities change as industries grow and as bigger cities use more fuel for space heating. The composition of the particulates changes, too, as industrial processes change, and sulfur pollution fluctuates with the sulfur content of the fuel burned. But the particulate-sulfur combination which made cities dirty, unpleasant, and unhealthy a hundred years ago and more was obviously still doing so during Episode 104.

Photochemical Smog. Just since World War II we have become aware of another kind of air pollution—*photochemical smog*. Particulates and sulfur dioxide are difficult enough to measure in the air, but it is impossible to obtain a daily average for the complex and fast-changing mixture of chemicals that is photochemical smog. Nevertheless, it was certainly one of the problems in Episode 104. The necessary ingredients were present during the episode—*hydrocarbons,* products of the unburned or partially burned gasoline from the exhausts of cars, trucks, buses, and from some industrial stacks; *nitrogen oxides,* produced from the natural nitrogen of the air by the heat of automobile engines, power plants, and other combustion; and *oxygen* in natural and plentiful supply. Sunlight, another necessary ingredient which must act on these constituents to produce smog, was also plentiful; with few exceptions, there were only scattered clouds throughout the area for the ten days of the episode.

Since hydrocarbons are one ingredient of smog, measurements of these chemical compounds could reveal whether or not smog levels rose during the episode. However, a large part of what is measured as total hydrocarbons is methane, some of which is produced naturally and some of which comes from leaks during the storage and transport of natural gas; methane does not react easily with other chemicals in the air as do the hydrocarbons from automobile exhausts. This means that the *nonmethane* hydrocarbons for peak hours were the significant data and these numbers were simply not available during Episode 104. Total hydrocarbons and methane were both measured in a few cities during the episode, but data sheets gave the maximum concentration in a single hour for each day—usually one hour for total hydrocarbons and a different hour for methane—making it impossible to compare nonmethane hydrocarbon peaks before and during the episode or with the present standard of 0.24 parts per million, maximum three-hour concentration. (Daily averages of total hydrocarbons did rise.)

Among the contaminants produced by photochemical reactions are *ozone* and *peroxyacetyl nitrate* (PAN), two important components of what appears on air pollution monitoring reports as *total oxidants.* Oxidants are another indicator of photochemical smog. Data were available for only four cities (one station each), but three of these recorded a rise over previous concentrations and, except in Akron, a rise above the federal standard during the episode:

Indicators of Photochemical Smog

	Nitrogen Dioxide (24-hour average in parts per million)		Oxidants (max. 1-hour average in parts per million)
City	Earlier Period in August*	Episode Average	Episode Peak
Akron	—	—	0.05
Chicago	0.053	0.087	0.12
Cincinnati	—	—	0.118
Cleveland	0.122**	0.132**	—
St. Louis	0.044	0.064	0.10
Federal Standard	—	0.05†	0.08

* Same days of the previous week; same number of days as in episode average.
** Average of ten stations, one day only.
† Standard is for annual average.

Oxidant levels in Chicago during Episode 104 showed no rise and even a slight drop in comparison with the days preceding the episode. However, the concentrations were high in Chicago for the entire three-week period of August 11–31, 1969, about double what they were in August of 1967. Measured or not, it is safe to assume that almost anywhere in the episode area where traffic was heavy there was some photochemical smog.

Carbon Monoxide. Carbon monoxide, another product of automobile exhaust, was measured at single stations in a few cities during Episode 104. Measurements of this pollutant are extremely sensitive to nearby traffic variations. Measurements inside cars on the street may be double the level recorded at the station, while in-traffic levels may be four times as high as the recorded level[10]. Carbon monoxide was monitored in Akron, Cincinnati, Chicago, and St. Louis, but only in the two latter cities was there a discernible rise. In St. Louis, the episode average was ten parts per million, while it had been 5.7 the previous week. In Chicago, the average rose from 8.5 to 11 parts per million. In St. Louis, carbon monoxide rose to 30 parts per million during one two-hour period and to 67 during one five-minute period, exceeding the federal standard that has since been adopted (nine parts per million not to be exceeded more than once per year).

Medical Warnings

As Episode 104 lengthened, some physicians became concerned. Dr. Bertram Carnow, Chief of Environmental Health at the University of Illinois Medical Center, issued a warning over Chicago radio stations suggesting that people with respiratory diseases remain indoors and limit their activity for the duration of the episode. Dr. Helen Bruce, acting City Health Commissioner in St. Louis, advised people to "hold down vigorous exercising" for the duration of the alert. Dr. John A. Pierce, Director of the Division of Pulmonary Disease at the Washington University School of Medicine and a member of the Committee for Environmental Information's Air Pollution Subcommittee, suggested that persons suffering from respiratory diseases should limit their movements in order to breathe as little of the polluted air as possible. He added that normal, healthy individuals could also expect to experience eye irritation, a feeling of fatigue, increased nasal secretion, and coughing. Although Chicago has provisions for restricting emissions when pollution reaches certain levels, these criteria were not met during this episode and therefore no action was taken. There was no emergency plan in St. Louis at that time, although one has since been adopted, but the city Air Pollution Control Commissioner said he would act to protect the city's people if dangerous pollution levels were reached[11].

On Wednesday, August 27, there were a few voluntary industrial cutbacks and persons holding temporary permits for open burning in the suburbs were ordered to stop such burning until the alert was over. By this time the smog was so thick in St. Louis that motorists were driving to work with their lights on and public viewing through the planetarium's telescope was canceled because the planets and stars were obscured.

Thursday began in St. Louis very much like the preceding three days: a low-level inversion (surface to 850 feet), an upper inversion based at 5200 feet, and no wind. But the high-pressure system that was responsible for the episode was drifting eastward, indicating that relief was on the way. It was just in time. The sulfur dioxide level was high at midday at all metropolitan area monitoring stations and so high all afternoon in the vicinity of a lead company, which is one of the area's most notorious polluters, that employees at the nearby Chart Center were dismissed early.

However, a surface breeze had begun to blow and the wind was also increasing higher up. The surface inversion dissipated and the base of the

St. Louis in the midst of Episode 104, with the industrial area on the Illinois side of the Mississippi River in the foreground and the river lost in smog in the middle distance. Two spans of one of the bridges are barely visible rising out of the smog. All that can be seen of the city is the arch and a cluster of the tallest buildings near it.

high inversion layer rose to 6000 feet. The episode was over in St. Louis on Thursday and in Chicago on Friday, and some communities that had begun to experience problems on August 22, had time to breathe in the middle of the following week before adverse conditions closed in again. For others, the HAPP advisory remained in effect for ten days from August 23 to September 1 without a letup. Before it lifted completely, Episode 104 had touched all or part of 22 states from the Great Lakes to the Gulf of Mexico and from west of the Mississippi to the Atlantic and had affected more than 20 million people[12].

Health Effects of Episode 104

There have been no studies of the health effects of Episode 104. Meteorological information can tell us when a high air pollution potential exists and ambient air monitoring can give us a general (although incomplete) idea of how high pollution levels are rising, but only the careful collection and analysis of relevant health statistics can tell us something of what an episode does to people. One consequence of this, as John Goldsmith[13] points out, is that "health effects are most likely to be found where trained personnel collect the right kind of information and undertake a careful analysis of it." Thus, Episode 29 in 1962 affected 18 major eastern cities for five to seven days, but only in New York do we know that during this period there was an increase in respiratory illnesses and deaths because only in New York did medical scientists look for these effects[14]. True, New York had the highest pollution levels and therefore we might expect the most serious health effects there, but the difference between New York and other cities is likely to be one of degree.

There were two important differences between Episode 104 and the famous air pollution disasters in which many have died. Such disasters occurred in the Meuse Valley in Belgium in 1903; in Donora, Pennsylvania, in 1948; in London in 1952, 1959, and 1962; and in New York in 1953, 1962, and 1966. In these disasters, measured or estimated concentrations of sulfur dioxide and particulates were higher for periods longer than any recorded in August of 1969. For example, Episode 73 during Thanksgiving in New York in 1966 was probably responsible for about 24 deaths a day for seven days[15], but the daily averages of sulfur dioxide then were 0.40 parts per million or more on two of those days, while this pollutant was recorded at that high a concentration for only a few hours during Episode 104. In the London disaster of 1952 in which about 4000 died, sulfur dioxide concentrations reached 1.34 parts per million, much higher than Episode 104 levels. However, the highest particulate level reported in that London episode was 446 micrograms per cubic meter[16], a level exceeded at one Chicago station during Episode 104.

The other difference between Episode 104 and these disasters was that in most cases the weather during the disasters included not only thermal inversions and calm or light winds, but also fog. Very small water droplets

The day Episode 73 (the "Thanksgiving Episode") broke up in New York. The top photograph was taken at 9:37 AM while smog still hovered over the city. The bottom photograph was taken at 1:06 PM the same day, from the same point.

can play an extremely important part in the chemistry of the atmosphere (in the transformation of sulfur dioxide to sulfuric acid mist, for example) and can also carry pollution deep into the lungs.

Without a careful study of the health effects of Episode 104 concentrations, we can only look at the health effects of similar concentrations observed elsewhere[17]. Scanty monitoring data make even this possible for only a few cities and such comparisons are far from exact.

During Episode 104, oxidant levels in St. Louis (0.07–0.08 in daylight hours) were high enough to cause asthma attacks in some chronic asthmatics. Construction workers or others engaged in hard physical work might have found it harder than usual to work in Cincinnati, St. Louis, and Chicago. Some impairment of athletic ability and other physical activity has been noted at the levels found in those cities. Eye irritation was also to be expected and, in fact, was a frequent complaint in St. Louis.

Carbon monoxide, if we accept the off-the-street measurements, did not reach a hazardous level anywhere except in St. Louis, and there only briefly. But when the St. Louis CAMP station recorded a two-hour average of 31 parts per million of carbon monoxide in the late afternoon of Monday, August 25, people driving down Twelfth Boulevard past the CAMP station may have been receiving as much as 60 parts per million while pedestrians could have been taking in as much as 120. More than a very brief exposure to 60 parts per million of carbon monoxide can cause acute distress to people with heart disease and impair both time and vision discrimination and cause headaches in healthy people. So there were probably some carbon monoxide effects in St. Louis, especially among smokers who received an added dose of carbon monoxide from their own personal sources.

Daily averages of sulfur dioxide at the St. Louis CAMP station ranged from 0.09–0.10 parts per million and particulate levels at the U.S. Post Office not far away were from 279–234 micrograms per cubic meter. Sulfur dioxide and particulates in these concentrations would be expected to produce increased respiratory symptoms in children and some adults, and more severe problems for people already suffering from chronic respiratory disease.

Daily averages of sulfur dioxide in the most polluted parts of Chicago were 0.08 and 0.09 parts per million accompanied by particulate levels in the same range as St. Louis, so similar effects on health would be expected. Particulate levels reached 300 micrograms per cubic meter at one Indianapolis station, but the sulfur dioxide level there remained uniformly low. In Toledo, on the other hand, sulfur dioxide went up to 0.11 parts per million, but particulates remained low at the only station reporting. A combination of high levels of both is usually considered to be more hazardous than a high level of either alone.

The most dangerous situation in St. Louis apparently developed on the final day of the episode, when sulfur dioxide levels rose higher and stayed up longer than on any of the previous days. Had the forecast not been for improved weather, St. Louis would have met criteria that now call for a second-stage or "yellow" alert, with some voluntary and some mandatory restrictions of emissions. Although the concentrations reached that day have been known, under similar circumstances, to cause a sharp rise in the illness rate of people 54 or older with severe bronchitis, a yellow alert will be called only if bad pollution weather is expected to continue for the next 24 hours.

But the full significance of the health effects of Episode 104 is not to be found in estimates of the immediate, acute effects—unpleasant and even serious as these may have been in many individual cases. Probably the most important health effect of the episode was that it delivered a *sustained* dose of higher-than-normal pollution to more than 20 million people. An analysis of Episode 29 in 1962 found that it contributed from one-fifth to more than one-half of the total pollution dose for November and December of that year. The report points out that "Generally this is the result of an increased *duration* of peak concentrations (a lengthening of the normal diurnal peaks) rather than striking increases in concentration."[18] In this way, Episode 104 contributed much more than a normal weekly dosage to the total pollution exposure in cities where there is always a measurable amount of pollution in the air.

The pall of dirty air had other effects too—effects impossible to quantify, but equally impossible to ignore. There was undoubtedly damage to vegetation and materials, the decreased visibility was an inconvenience and a hazard to drivers, and there were the intangible, depressing effects of the masking of the stars at night and the sun in the daytime, the unpleasant smell, and the reluctance to take a deep breath.

Episode 104 also delivered an unknown dose of pollution to people in small towns and rural areas where there was no pollution monitoring that week and where little or no pollution monitoring is ever done. The episode was not just an intensification of urban pollution. It was a reminder that air pollution has long since ceased to be a local, big-city problem.

Beyond the Cities

Except for the information produced by one monitoring station in Crystal City, Missouri, a small town 35 miles south of St. Louis, no monitoring data were available outside the metropolitan areas involved in Episode 104. Another special station in St. Charles, although outside the central city, was in a town of 21,000 which has recently been engulfed by the westward expansion of the metropolitan St. Louis area.

At St. Charles, the coefficient of haze *(coh)*—one way of measuring dirty air—was comparable to downtown St. Louis, while the highest coh readings in Crystal City were equal to the lowest episode readings at the St. Louis CAMP station. Sulfur dioxide levels at Crystal City were low, but at St. Charles they did equal St. Louis daily averages for a four-hour period on the last day of the episode. However, without other measurements of the same pollutants from the same stations for comparison, it is impossible to know how these measurements related to the usual air quality in these towns. From other Missouri state data[19], it is now clear that sulfur oxide pollution in the city of St. Charles and other parts of St. Charles County is increasing.

One of the voluntary measures taken to keep Episode 104 pollution from becoming more acute in St. Louis was Union Electric's shift of some of its power production to outlying plants. One of these is in St. Charles County and another is ten miles south of the city. Union Electric, the utility serving the St. Louis metropolitan area, has grown from a capacity of 721,000 kilowatts prior to 1940 to two million in 1960 and to over four million in 1969. By 1973, the utility will add another 2,400,000 kilowatts to its capacity. All the new and projected plants are well outside the city—a fact that has been emphasized by utility spokesmen in urging that the new plants be exempt from standards for the emission of sulfur dioxide.

The spreading of power plants to the periphery of metropolitan areas has now become a national trend and, while sulfur oxide and nitrogen oxide concentrations in the cities may no longer increase proportionally to the increase in power, power-plant pollution is spreading over a wider area. Not only power plants but many other industries are now located outside of metropolitan areas—sometimes very far from the city limits—and in such locations monitoring and control of air pollution are usually poor and often nonexistent. It has been suggested that this is a valid principle in urban growth planning. Using Chicago as an example, the authors of a recent paper[20] showed how urban sulfur oxide pollution could be reduced in the city if utilities and manufacturing were moved out. However, any improvement in city air made in this way is achieved at the expense of deteriorating country air.

Although there was little data specifically pertaining to Episode 104 from the small towns and none from the rural areas affected, the increasing spread of city smog and dirt to outlying areas has been documented in a number of ways. A series of papers based on the collection and analysis of particulate samples in the National Air Surveillance Network[21] show that since 1958 there has been what is described as "an unrelenting upward slope" in the concentration of particulates in nonurban samples. Such typical urban constituents as lead are now found at nonurban stations near cities, in lesser amounts at rural stations, and in measurable, though still smaller, amounts even in very remote locations. Robert V. Bega, director of forest disease research at the University of California at Berkeley, has found smog damage to trees as far away as 114 miles east of Los Angeles in the Palm Springs area[22]. Raymond F. Falconer of the Atmospheric Sciences Research Center in Wilmington, New York, has found increasing particulate contamination at the Center, 120 miles from Albany and 260 miles from New York City[23].

Coming Events

Obviously, inadequate data have prevented us from achieving a real understanding of what was happening in the cities during Episode 104, and we know still less about the situations in suburban and rural areas. But people who experienced the episode need no technical data to tell them that a repetition is undesirable. Nevertheless, meteorology tells us that a repetition of high air pollution *potential* is inevitable.

Some cities are now hoping to prevent episodes from turning into disasters by employing various emergency plans to curb automobile use, waste incineration, power generation, and industry before pollution levels become high enough to cause deaths. No disaster plan has yet gone into effect and no one knows whether such a plan could successfully hold pollution below the danger level. It is worth noting that Episode 73 was a killer in New York City in 1966, in spite of the fact that it occurred when temperatures were moderate—too cool to require power for air conditioning and too warm to require much power for space heating. (Episode 73 also occurred on a holiday weekend, which automatically produced cuts in both traffic and industry.) None of the emergency plans would affect an episode like 104 since they do not call for measures to reduce pollution until it becomes just a little worse than it was in August, 1969, or until it lasts just a little longer.

To know how often to expect a repetition of Episode 104, or something much like it, we must therefore turn to the meteorologists. The number 104 gives us part of the answer. This was the 104th episode of high air pollution potential since August 1, 1960, covering 75,000 square miles or more and lasting more than 36 hours[24]. (Episodes west of the Rocky Mountains were not included until October 1, 1963.) High air pollution potential was frequently experienced in areas smaller than 75,000 square miles or for periods of less than 36 hours. These numerous local episodes are not numbered, but, according to Morris Neiburger, a meteorologist at the University of California at Los Angeles, "Inversions occur more than one-fourth of the time over almost all of the United States. Anywhere pollutants are emitted into the air, high concentrations can be expected a considerable fraction of the time."[25]

The Environmental Science Services Administration (ESSA) map shows the total number of days various areas of the country have been included in

The number of days that widespread high air pollution potential was forecast throughout the United States are shown in this map from the Environmental Science Services Administration. The program began in the East on August 1, 1960, and in the West on October 1, 1963. Between these respective dates and April 3, 1970, there were 39 episodes in the West and 75 in the East. The numbers indicate the days a particular area was affected by a HAPP forecast. For example, the area between the line marked 0 and the line marked 10 was affected between 0 and 10 days. This map does not show the many additional days of bad air pollution weather of less than 75,000 square miles extent, or of less than 36 hours duration, or both.

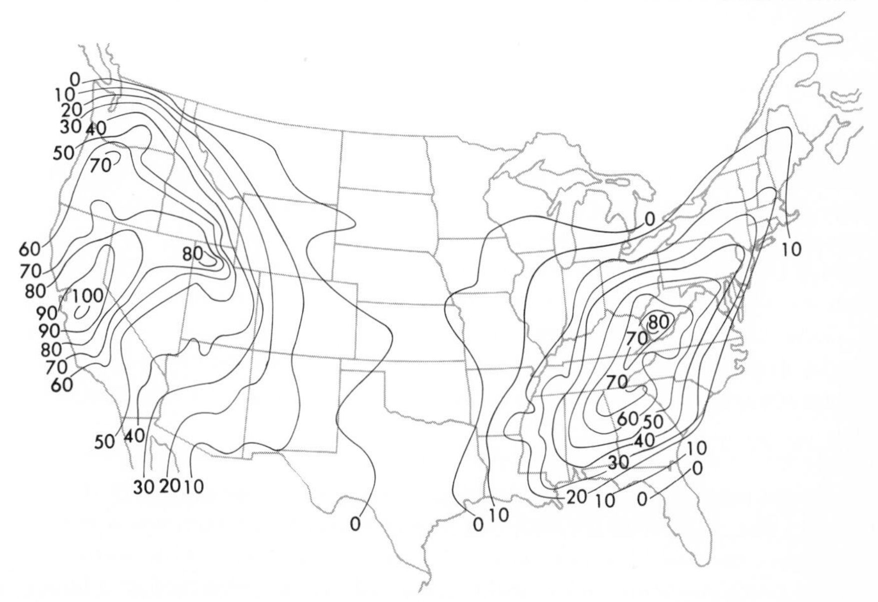

widespread episodes since the HAPP program went into operation. Only a strip down the center of the country through the Plains states has been exempt. The map shows that pockets in both the East and the West are affected exceedingly often. These very pockets also suffer most from shorter and more local periods of high air pollution. Los Angeles in the West and the West Virginia Kanawha Valley (Charleston and vicinity) in the East are examples: Los Angeles is under inversions more than half the time during winter months[26]; shallow inversions have been found at sunrise in the Kanawha Valley 74 percent of the time in winter, 78 percent of the time in summer, and 92 percent of the time during spring and fall[27].

As is well known, the dominant air pollution problem in Los Angeles is photochemical smog, partly because industrial emissions are better controlled there than elsewhere (making automobiles the major pollution culprits) and partly because of the preponderance of sunny days. On the same day Episode 104 began in the Midwest, Los Angeles had its second smog alert of the year, with an ozone concentration of 0.7 parts per million[28]. A second-stage alert, calling for restrictions of emissions, has not yet been called in Los Angeles, but the city has had from one to six first-stage alerts every year thoughout the sixties[29], and has failed to meet the California oxidant standards about 200 times yearly[30]. The Kanawha Valley is a center of chemical industry and has 11 steam-generating power plants. A 1964–1965 study of the valley's air pollution revealed particulate concentrations of 272 micrograms per cubic meter of air—*not* during an episode, but as an *annual average.* Daily averages of sulfur dioxide were higher 99 percent of the time than the daily episode levels in St. Louis[31]. Clearly both communities had high air pollution potential on more days than the 50 shown on the preceding map for Los Angeles (1963–1970) and the 80 shown for the Kanawha Valley (1960–1970).

Areas with many bad days should not be the only causes for concern, however. The New England area has had only ten HAPP days since 1960, but six of these were during the severe Episode 29 in 1962. This episode covered some of the most densely populated areas of the country—"areas not usually subject to frequent or long stagnations," as the Public Health Service report[32] pointed out. Based on an analysis of stagnating anticyclones for the period of 1936–1956 which had found that the areas most affected during November–December 1962 had experienced only two or three such long stagnations during those years, the report predicted that "an episode of this magnitude can be expected in these areas about once every ten years."[32]

"Inevitable" Episodes

Air pollution episodes that may develop into disasters appear inevitable only because our technological civilization continues to spew thousands of tons of pollutants into the air every day. If it were not for the smokestacks and car exhausts, the migratory high-pressure system that caused Episode

Episode 104 in the Kanawha Valley of West Virginia. This is one of the most unfortunate areas of the country for bad air pollution weather. Very large amounts of pollution are emitted in Charleston and vicinity by chemical plants and other industry.

104 would have drifted across the continent, producing only a series of warm, lazy August days with clear skies, good enough for swimming in the daytime and stargazing at night. That we dread such weather phenomena is an outgrowth of our technology and of our present approach to air pollution control.

The guiding principle for the air pollution control efforts of the sixties has been stated as *optimum use of our air resources*[33]: The air can be used by everyone for breathing as long as pollutants do not build up to unsafe concentrations. It can be used by power plants, industry, automobiles, trucks, buses, and planes to make possible the combustion that is an essential part of power production, manufacturing processes, and transportation. It can also be used as a medium for the disposal of the waste products from all these sources up to the limit of the atmosphere's capacity for dilution and dispersion. Only when this capacity is exceeded—when wind and weather can no longer protect people—are additional controls required. In practice, there has often been no real plan aimed at achieving a particular air quality. Rather, a community's control efforts have been determined by balancing public pressure and industrial resistance, technical feasibility and political feasibility, health considerations and budgetary considerations.

Both the guiding principle and the actual practice of "optimum use" have serious weaknesses, as Episode 104 sharply revealed. First, polluters are permitted to continue loading the atmosphere over the cities *as long as weather conditions are favorable;* in an episode, the atmosphere's capacity to dilute and disperse these pollutants simply breaks down. Second, polluters are permitted to use the atmosphere *outside urban centers* for waste disposal almost without restraint; the effects of this type of pollution are then rarely noticed until an outlying area is caught up in urban growth or until an episode makes it clear to our eyes and noses, even in the absence of monitoring, that it is a remote area indeed that still has truly clean air. While some cities are beginning to reduce the emission of some pollutants, these gains are being overwhelmed by the growth of power, industry, and traffic. The spread of some pollutants, carried by air currents, has already become worldwide.

The story of Episode 104 shows that two factors must combine to create a pollution episode—unfavorable weather conditions and pollution emissions. We cannot change the former. Until we drastically reduce the latter, episodes will continue to occur; some will affect thousands, some will affect millions, and some will turn into disasters.

Episode 104 also emphasized the need to look beyond the cities. While the concern with air pollution has broadened to the recognition that it is a regional and not simply an urban problem, there is not yet sufficient understanding that pollutants never just "vanish into thin air." They may remain airborne, becoming part of the suspended particulates of a suburban or rural area, of the air thousands of miles away from the city that produced them, or of the atmosphere high above the earth. They may eventually fall to the ground or into the water—a fate that does not necessarily render them harmless. As gases in the form they were emitted, or chemically changed, pollutants may affect the composition of the earth's envelope of air and consequently its climate. Pollutants can do permanent damage to human health, to the atmosphere itself or—when they reach water, soil, or plants—to some other aspect of our ecosystems. The first step in the direction of understanding what air pollution is doing to people and to their environment is to understand the atmosphere in its natural state and the role it plays in the lives of people and in the living world—the biosphere.

Acknowledgments

Data from the various cities were kindly provided by the following departments and individuals: Department of Environmental Control, Chicago, Illinois; City Pollution Control Agency, Toledo, Ohio; Charles G. Beard, Executive Director, West Virginia Air Pollution Control Commission and Charles R. Lewis, Executive Assistant, West Virginia State Medical Association; John L. Calder, III, Air Pollution Control Bureau, Hamilton County, Tennessee; John E. Clausheide, Chief of Air Pollution Control, Evansville, Indiana; Charles Copley, Air Pollution Control Commissioner, City of St. Louis; George T. Craig, Chief, Bureau of Technical Services, Cleveland Division of Air Pollution Control; Jerry J. Garro, Air Pollution Technician, Akron, Ohio; Kenneth Irwin, Air Monitoring Services, Air Pollution Control District of Jefferson County, Kentucky; David Lee, Pollution Control Specialist, Air and Water Pollution Control Commission, State of Mississippi; Donald Pecsok, Director, Division of Air Pollution Control, St. Louis County Health Department; Charles B. Robison, Assistant Director, Bureau of Environmental Health, Jefferson County, Alabama; Charles E. Schumann, Air Pollution Chemist, Division of Air Pollution Control, Metropolitan Sewer District of Greater Cincinnati; Anthony Telford, Supervisor, Environmental Protection Agency, Madison, Monroe, and St. Clair Counties, Illinois; R. E. Wetzel, Superintendent, Bureau of Air Pollution Control, Indianapolis, Indiana.

[1] An article by John R. Stallard in the *Milwaukee Journal,* August 20, 1970, p. 4, lists the violators. Milwaukee had no standards for other pollutants at that time.

[2] *Report of the St. Louis Area Air Pollution Technical Coordinating Committee on the Air Pollution Episode in the Metropolitan St. Louis Area, August 25–28,* Appendix I, 1969, p. 2.

[3] Environmental Science Services Administration, Local Climatological Data, St. Louis, Missouri, August 1969. Reports from stations at Lambert Field and Jefferson National Expansion Memorial.

[4] Morris Neiburger, "The Role of Meteorology in the Study and Control of Air Pollution," *Bulletin of the American Meteorological Society,* Vol. 50, No. 12, December 1969:960. (This article does not discuss Episode 104 in particular, but reports data from studies of Los Angeles episodes.)

[5] Monitoring in Alabama has since been resumed. In 1971, Birmingham experienced an episode during which particulate concentrations exceeded the highest recorded anywhere in Episode 104: a record 607 micrograms (millionths of a gram) per cubic meter of air on Monday, April 20. The count had been up to 520 on the previous Thursday and 499 on Friday. A HAPP warning was received on Friday, but although the readings were 200 on Saturday and 283 on Sunday, control officers felt that "local forecast and ambient air

quality over the weekend period did not indicate an air pollution problem of increasing significance." (Charles B. Robison, Assistant Director, Bureau of Environmental Health, Jefferson County Department of Health, personal communication, August 13, 1971.)

[6] Gladwin Hill, "Air Pollution Grows Despite Rising Public Alarm," the *New York Times,* October 19, 1969, p. 61. (NAPCA has since been replaced by the Office of Air Programs in the new Environmental Protection Agency.)

[7] Thomas B. McMullen, "Comparison of Urban and Nonurban Air Quality," presented at the Ninth Annual Indiana Air Pollution Control Conference, Purdue University, October 13–14, 1970

[8] Allen A. Nadler et al., *Air Pollution,* Scientists' Institute for Public Information Workbook, New York, 1970, p. 19.

[9] D. A. Lynn, B. J. Steigerwald, and J. H. Ludwig, *The November–December 1962 Air Pollution Episode in the Eastern United States,* U.S. Department of Health, Education, and Welfare, Division of Air Pollution Control, Cincinnati, Ohio, September 1964, pp. 8–9.

[10] J. D. Williams et al., "A Proposal for an Air Resource Management Program," Vol. VIII of the National Center for Air Pollution Control's *Interstate Air Pollution Study: Phase II Project Report,* Cincinnati, Ohio, May 1967, p. 66.

[11] The *St. Louis Post-Dispatch.* Charles Copley, Air Pollution Control Commissioner, was quoted on August 26, 1969; Dr. Bruce and Dr. Pierce on August 28, 1969.

[12] The total population of the metropolitan areas from which some episode information was available is about 20 million. This omits all cities with a population of less than 100,000 and some larger ones (such as Nashville and Memphis) which did not respond to a request for information.

[13] John R. Goldsmith, "Effects of Air Pollution on Human Health," *Air Pollution,* Vol. I, Arthur C. Stern (ed.) (New York and London: Academic Press, 1968), p. 560.

[14] J. McCarroll and W. Bradley, "Excess Mortality as an Indicator of Health Effects of Air Pollution," *American Journal of Public Health,* 1962, 56:1933–42.

[15] M. Glasser et al., "Mortality and Morbidity During a Period of High Levels of Air Pollution, New York, November 1966," *Archives of Environmental Health,* 1967, 15:688.

[16] Goldsmith, "Effects of Air Pollution on Human Health," p. 558 (see [13]).

[17] All health effects in the remainder of this section are taken from Air Quality Criteria documents of the National Air Pollution Control Administration, Washington, D.C., as follows: *Particulate Matter,* AP-49, January 1969; *Sulfur Oxides,* AP-50, January 1969; *Carbon Monoxide,* AP-62, March 1970; *Photochemical Oxidants,* AP-63, March 1970; *Hydrocarbons,* AP-64, March 1970.

[18] Lynn, *The November–December 1962 Air Pollution Episode in the Eastern United States,* p. 19 (see [9]). Emphasis added.

[19] Missouri Air Conservation Commission, *Missouri Air Quality Data,* biannual reports beginning January–June, 1968.

[20] Alan S. Cohen and Arthur P. Hurter, "Urban Evolution and Air Pollution," presented at the 63rd Annual Meeting of the Air Pollution Control Association, St. Louis, Missouri, June 14–18, 1970.

[21] John H. Ludwig, George B. Morgan, and Thomas B. McMullen, "Trends in Urban Air Quality," presented at the national fall meeting of the American Geophysical Union, San Francisco, December 17, 1969. Also Thomas B. McMullen, R. B. Faoro, and George B. Morgan, "Profile of Pollutant Fractions in Nonurban Suspended Particulate Matter," *Journal of the Air Pollution Control Association,* Vol. 20, No. 6, June 1970: 369–72.

[22] UPI story in the *New York Times,* December 12, 1969.

[23] The *New York Times,* May 24, 1970.

[24] For technical details of the HAPP criteria, see *ESSA Technical Memorandum* WBTM NMC 47, U.S. Department of Commerce, National Meteorological Center, Washington, D.C., May 1970.

[25] Neiburger, "The Role of Meteorology in the Study and Control of Air Pollution," p. 959 (see [4]).

[26] Charles R. Hosler, "Low-Level Inversion Frequency in the Contiguous United States," *Monthly Weather Review,* September 1961, 89: 319–39.

[27] National Air Pollution Control Administration, *Kanawha Valley Air Pollution Study,* Raleigh, North Carolina, March 1970, p. 2–31.

[28] Dick Main, "Smog, Record Heat Plagues Southland," the *Los Angeles Times,* August 22, 1969.

[29] Air Pollution Control District, Los Angeles County, *Profile of Air Pollution in Los Angeles County,* January 1969, p. 64.

[30] Lowell Wayne, Vice President and Director of Research, Pacific Environmental Services, Inc. (personal communication).

[31] *Kanawha Valley Air Pollution Study,* pp. 5-11 and 5-22 (see [27]).

[32] Lynn, *The November–December 1962 Air Pollution Episode in the Eastern United States,* p. 6 (see [9]). The study of anticyclones–which this prediction is based is J. Korshaver, "Synoptic Climatology of Stagnating Anticyclones," SEC Technical Report A60-7, Robert A. Taft Sanitary Engineering Center, Cincinnati, Ohio, 1960.

[33] *Interstate Air Pollution Study: Phase II Project Report,* Vol. VIII, p. 6 (see [10]).

The Earth-Atmosphere System

Man's headlong rush to conquer and exploit nature has fed on its own success and has encouraged him to assume that this process will continue indefinitely—that he will be able to achieve ever greater successes in adapting the resources and processes of the earth-atmosphere system to his own purposes. At the same time, the human scale is so puny when set against the immensity of nature—the wealth of energy pouring forth continuously from the sun, the seemingly boundless atmosphere, the vast, slow-moving elemental cycles, the intricately interwoven systems of living things—that it has appeared inconceivable that man could have more than a local and temporary effect on nature or interfere in any significant way with its continued ability to support life—including his own.

In recent years, the acceleration of technological change has forced us to question both of these basic assumptions and to recognize that both arose from ignorance. Still ignorant of the world we live in and the manner and extent to which we can safely use it, we are just learning to ask some of the right questions and to get some uncertain and incomplete answers.

Very Special Conditions

Our earth is indeed a rich and fortunate planet, imbedded in the universe in such a way that its size, its position, its orbit, and its rotation combine to provide the optimum conditions for life: a protective atmosphere, life-giving water, neither too much nor too little solar radiation.

If the earth were much smaller like the moon, its gravitational force

would be too weak to hold an atmosphere near its surface. The origin of much of the earth's present atmosphere was probably outgassing from the interior of the earth, but even if volcanoes on the moon emitted the same combination of gases as the earth's volcanoes, these gases would not remain near the moon's surface, but would dissipate into space. If, on the other hand, the earth were much larger like Jupiter, its gravitational force would be so great that it might still have the primordial atmosphere of quite different composition which is thought to have surrounded the earth billions of years ago and still surrounds Jupiter and some of the other planets today. (Jupiter's atmosphere is so dense that near that planet's surface a human being would be crushed.) As it is, the earth probably lost its primary atmosphere in the hot days of its youth and slowly developed its own peculiar atmosphere. Its constituents are constantly being removed and constantly renewed in the great cyclical processes by which the earth's atmosphere reacts with its oceans and continents.

The earth and the atmosphere which surrounds it form an almost closed system. In the neighborhood of its outer edge, the atmosphere becomes so hot that molecules of hydrogen and helium do move fast enough to overcome the gravitational pull of the earth and escape into space. But with these miniscule exceptions, every material that goes into the atmosphere—gas, liquid, or solid—may be moved around, may change its form, may be deposited in the oceans or on the land, *but remains within the system.* Although we may not be able to trace the pathways of these materials or account for the quantities of them that remain in the air at any given time,

The influences which determine the characteristics of the atmosphere extend millions of miles out in space; yet that part of the atmosphere in which living things can breathe (the lower part of the troposphere) is only a thin shell, surrounding another thin shell—the soil and water of the earth's crust. Thus, although earth is embedded in the universe in such a way that optimum conditions are provided for life, the biosphere in which life came into being and is sustained is contained within a very narrow band at the earth's surface. This satellite view of earth surrounded by swirling clouds suggests the limits of the earth-atmosphere system.

we can be assured that they have not "gone away." As the sun's energy pours into this system from the outside, its interaction with these materials and with the earth itself accounts for the amount of solar energy that can actually be used by the system.

As an envelope which encompasses the earth, its atmosphere is finite, with 99 percent of its mass within 20 miles of the earth's surface. As this atmosphere moves around the earth, however, it has no beginning and no end. The air over New York is part of the same atmosphere that covers Tokyo and Buenos Aires. The air we breathe at an ocean beach is part of the same atmosphere that is outside the window of a jet aircraft traveling six miles above, that is 20 miles above in the stratosphere, and that is 40 miles above in the mesosphere. For although these atmospheric layers are separated by temperature barriers (shown opposite), these barriers do not separate in any absolute sense and all atmospheric layers interact.

The atmosphere's inner edge or boundary layer is always and everywhere in contact with the land and the waters and the living things that inhabit them. The atmosphere as we know it and breathe it was created in part by living things; its present composition is maintained by them and, at the same time, all life is dependent upon the atmosphere.

Because of its oneness and its intimate interrelationships with land, water, and life, a change in the atmosphere at one point can be expected to affect some aspect of the earth-atmosphere system at some other point. This may be a simple, linear (one-to-one) effect on some small part of the system

The changes in temperature with altitude are shown in this schematic representation of atmospheric temperature. Pressure is indicated on the right. Above the protective lower layers, the temperature rises steeply. The *tropopause*, the boundary between the troposphere and the stratosphere, is not continuous at a single altitude, but is highest over the equator and lowest over the poles.

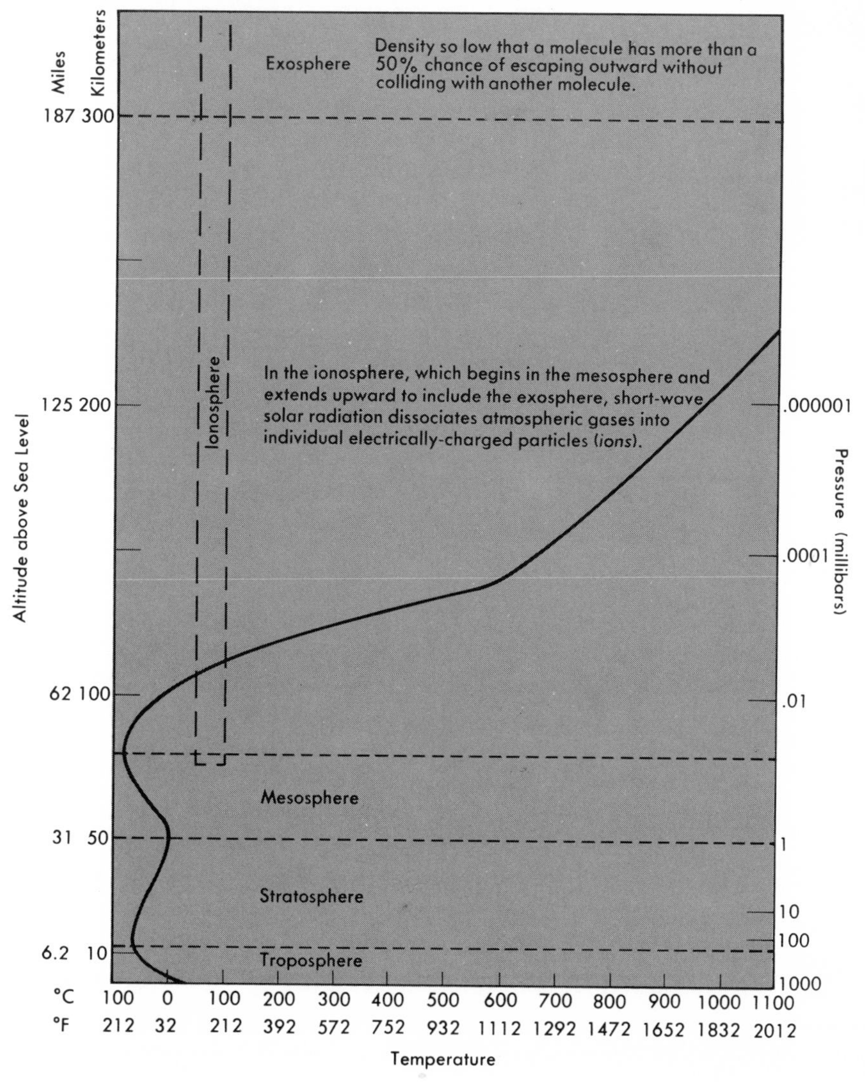

or a multiplying effect with one interaction triggering another. A compensating change balancing the first may be fed back into the system, or the initial effect may reverberate throughout the system in ways almost possible to trace. Because of all these possible complex and dynamic interrelationships, the whole is quite different and vastly greater than the sum of its parts. In spite of the immensity of the forces acting on the earth, very small portions of these forces and very delicate balances among them can actually be decisive in the operation of the system and in the maintenance of life on earth. It is for this reason that human activities, in spite of their puny scale, are capable of serious interference with the system which makes both humans and their activities possible.

Radiation Balance

The earth is about 93 million miles from the sun and travels around it in what might be called a "life belt": far enough from the destructive radiation that gives Mercury—36 million miles from the sun—its burned and blackened surface, yet close enough to it to escape the cold fate of Jupiter—483 million miles from the sun.

The sun's radiation pours down day after day and month after month, with what finally reaches any part of the lower atmosphere and the earth's surface varying with the time of day, the season, the latitude, the cloud cover, the changing content of carbon dioxide, water vapor, and particulates in the air above, and the reflective properties of the surface. The comparative values which follow suggest the immensity of the solar energy we receive on the earth. The total energies of certain phenomena and their atmospheric processes relative to the total solar energy intercepted by the earth each day* are:

Solar energy received on earth per day	1
Melting of average winter snow in spring	0.1
World use of energy in 1950	0.01
Average cyclone	0.001
Average hurricane	0.0001
Detonation of thermonuclear weapon in April 1954	0.00001

The earth itself absorbs a little more than half the incoming solar radiation and, in turn, radiates mostly invisible heat waves in the long-wave infrared part of the spectrum. Over a year, over the whole extent of the earth and its atmosphere, each incoming unit of solar radiation reaching the top of the atmosphere is balanced by an equivalent unit of outgoing radiation.

On a diurnal or even a seasonal time scale, or looking at only a portion of the earth's surface rather than at the whole globe, however, incoming and

* Adapted from William D. Sellers, *Physical Climatology* (Chicago: The University of Chicago Press, 1965), Table 3, p. 13.

outgoing radiation are *not* in balance. It is the relatively small differences between these two processes that are decisive in the operation of the system. This is why weather is ever-changing and why, in spite of the immensity of the solar radiation the earth receives, a comparatively small change in some aspect of either the incoming or the outgoing radiation could eventually effect a large change in the system.

Such a change can take place at many different points between the arrival of solar radiation in the lower layers of the atmosphere and the departure of terrestrial radiation into outer space. As radiation enters the atmosphere, some is simply reflected back into space. Condensed droplets of water or ice crystals reflect so well that, on a day when the whole sky is covered with clouds, more than half the incoming radiation is reflected in contrast to only about 15 percent on a clear day. The fine particles that remain suspended for a few days (in the lower layers of the atmosphere) to more than a year (higher up) scatter the sun's radiation in all directions—some down toward the earth, some back toward space. The earth's surface itself has varied reflecting properties; snow fields are the most effective reflectors, while green forests reflect the least.

The gases and particles in the atmosphere absorb about 16 percent of the solar radiation on a clear day. In particular, water vapor and carbon dioxide (practically transparent to visible light, but not to invisible heat) absorb some of the sun's infrared radiation, thus protecting the earth from excessively high daytime temperatures, but allowing the visible light to pass through. This leaves some 69 percent of solar radiation to be absorbed at the surface on a clear day and only 35 percent to be absorbed on a cloudy day. On land, the absorbing layer that shows seasonal temperature changes may be from 16 to 66 feet deep; in the oceans, with their constant motion and mixing, it may be from 160 to almost 2,000 feet deep. The oceans therefore store much more of the sun's radiation as heat than the land does, which explains why coastal climates are moderate while landlocked areas in the same latitude band are much hotter in summer and much colder in winter.

After the sun goes down, the earth's radiation predominates. Now the water vapor and carbon dioxide protect the earth from cooling too rapidly by reflecting heat back to the earth (remember, heat waves cannot pass through them). If the night sky is overcast, the warming effect is greater because the clouds absorb the heat and reradiate it downward.

Dense collections of particles may cool the lower atmosphere by reflecting solar radiation upward before it reaches the earth or may warm the atmosphere by absorbing and reradiating heat either from the earth or from the sun. Whether the cooling or heating effect is dominant depends upon the size and quantity of the particles, their composition, their location, and their interaction with atmospheric water vapor and other gases. Particles may attract moisture, forming the nucleus of a cloud droplet so that their presence strongly affects cloud formation and adds further to the upward reflection of solar radiation.

All of these many radiation fluxes tend to balance one another and to maintain an annual average temperature over the globe that fluctuates only slightly from year to year. As long as there is no annual change in either the number of incoming energy units of solar radiation or in the average distribution of each unit among the various absorptive and reflective processes (the interactions between the sun and the matter in the system), there will be no basic change in *global* climate despite local weather variations. We cannot be certain that the incoming solar radiation—the *solar constant*—is really unchanging, but at present scientists think it varies little, if at all [1]. The *distribution* of this radiation within the earth-atmosphere system may actually change, but because there is already so much variation in radiation from time to time and from place to place, widespread changes in climate are hard to detect except as they are revealed in changes in mean annual temperature. If more radiation is absorbed by the earth, oceans, and atmosphere this year than last year, without a compensating change in the radiation that is given off by the earth-atmosphere system, the mean annual global temperature will rise a fraction of a degree; if the balance is tipped in the other direction, the mean annual global temperature will drop.

The mean annual temperature in both hemispheres has changed perceptibly in the last century, as shown in the opposite graph. For about 60 years, beginning in the 1880s, the average temperature of the Northern Hemisphere rose 0.5° C. Since the mid-forties, this temperature has been dropping, although it has not yet returned to the 1880 value[2].

On a longer time scale, we know there have been much larger perturbations, for geological records indicate that during most of the earth's life northern temperatures have been ten degrees warmer than they are today and both poles have been free of ice. In the colder intervals, however, glaciers have sometimes extended over much of the earth's surface. During the last ten thousand years, although there has been no worldwide glaciation, the polar ice caps have not melted, so it is probably more appropriate to refer to our present age as "interglacial" rather than to assume that it is postglacial.

Once an ice age has begun, it is not difficult to see how the glaciers might get thicker and push farther and farther south from the Arctic. For example, as the temperature drops, more precipitation is in the form of snow, with the snow cover remaining on the ground longer. Snow reflects more of the sun's radiation than other surfaces, thus further lowering the temperature. (The reverse, of course, is also true: As the temperature rises, less precipitation is in the form of snow and less of the sun's radiation is reflected, thus further raising the temperature.)

But what *initiates* an ice age? How much of an imbalance between incoming and outgoing radiation must there be? How is a new balance achieved so that a glacial epoch can last for a million years or more? Within a glacial epoch what produces the interglacial ages such as our own when the ice retreats and a mild climate spreads in its wake? Is the triggering mechanism a change in the amount of solar radiation reaching the earth's

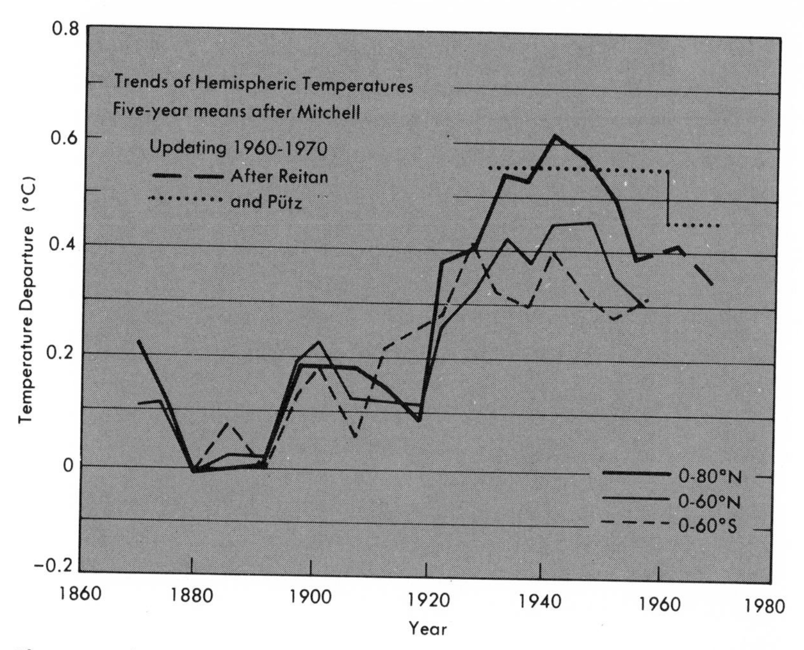

After rising for about 60 years, the mean annual temperature of the Northern Hemisphere is now falling, though it is not yet back to its 1880 value. Note that the peak reached around 1940 is much higher when the far northern latitudes are included. In the Southern Hemisphere, temperature changes in the latitude band between the equator and 60° South have been quite similiar to those between the equator and 60° North.

atmosphere or a variation in the earth's orbit around the sun? Could it be changes in the earth's crust—mountain building, volcanic activity, variations in sea level, continental drift—or a change in the carbon dioxide content of the atmosphere? Or perhaps some combination of two or more of these?

These questions have attracted the attention of scientists for many years and have resulted in a number of ice-age theories. Which one of these theories best describes reality is of more than academic interest as we cast an uneasy glance over our shoulders, questioning the stability of our present interglacial period and wondering whether our activities could inadvertently trigger either a new ice age or a warming trend of such magnitude as to melt the polar ice caps.

What *is* clear is that the processes that regulate our climate are complex and interrelated. A change initiated in one of these processes may be compensated for elsewhere, thereby maintaining the system's equilibrium. However, we do not know the extent of the system's ability to make these adjustments. A change may instead gather momentum and prove very difficult to reverse. Since these processes depend in part upon the material with which the solar radiation interacts, preserving the same overall atmospheric composition is important in maintaining the same global climate from year to year.

Composition of the Atmosphere

To describe the composition of the atmosphere is to describe the distribution of matter within it. Although this matter can be seen only when clouds, plumes of smoke, or layers of dust or photochemical smog are formed, even air that is pure and clean is not the nothingness that it appears to be; rather, it is composed of molecules of various gases of known ratio. This ratio is maintained by constant change, constant removal and replenishment, and constant interaction of the gases. The energy for all this activity is supplied by the sun.

Nitrogen dominates the atmosphere quantitatively; in dry air, it is about 78 percent by volume. Ammonia, a gas composed of nitrogen and hydrogen, was probably one constituent of volcanic eruptions and other outgassing from below the earth's surface many millions of years ago. It is thought that the nitrogen atoms in ammonia were split off by the action of sunlight and lightning, and the nitrogen in the atmosphere gradually built up to its present abundance.

The second most abundant constituent of the earth's atmosphere is oxygen (about 21 percent). Oxygen is far more reactive chemically than nitrogen, combining much more readily with other molecules and playing a key role in almost all life processes. Life, in turn, played a key role in the development of the oxygen-rich atmosphere. Although there was water vapor in the early atmosphere and each molecule of this gas contained an atom of oxygen and two atoms of hydrogen, there was no free oxygen gas consisting of molecules which united two oxygen atoms and were not combined with any other element. Ammonia and methane (a compound of carbon and hydrogen) were both components of the early atmosphere. With energy from the sun, everything necessary to create organic compounds—the food of every living thing—was present. (Laboratory experiments have shown that energy from ultraviolet light, an electric spark, or heat will produce organic compounds from an ammonia-methane-water mixture.)

As more and more of these compounds developed, it has been suggested that the oceans became a kind of "organic soup."[3] It was within this organic soup that life first appeared and the long evolutionary process of natural selection began—a process which was to lead to the development of forms of life capable of producing that oxygen so necessary to further evolutionary development. The first forms of life were almost certainly *anaerobic* (not requiring oxygen), and the first metabolism was the anaerobic metabolism typified by fermentation—still an important part of human physiology as well as the physiology of other life forms today. Some of these anaerobic life forms developed a form of photosynthesis, a mechanism for absorbing light and using it to synthesize organic compounds. Finally, green-plant photosynthesis developed, with green plants taking in energy from the sun and using it to split the oxygen atoms away from the hydrogen atoms in water molecules, thus making free oxygen possible. This green-plant photosynthesis was responsible for our present oxygen-rich atmosphere and still maintains it. At the same time, the presence of free oxygen made possible the proliferation of the myriad forms of *aerobic* (oxygen-requiring) forms of life that inhabit this planet today, including man.

In this view of the origin of life, the oxygen content of the atmosphere, the oceans, and the surface waters in past millenia is seen as slowly increasing, replacing hydrogen which escaped into space, replacing carbon which was removed and stored, and thus building the enormous reserves of oxygen in our present biosphere. Finally, the oxygen being taken in by bacteria, insects, fish, birds, mammals—all the aerobic forms of life—reached an equilibrium with the oxygen being returned to the atmosphere by the green plants. The oxygen content of the atmosphere then ceased to increase and has remained in an overall steady state—probably for millions of years.

Oxygen was—and still is—important in developing and maintaining the protection the atmosphere provides for the earth and its life below. Since the earth and the moon are about the same distance from the sun, the earth's temperatures would be like those of the moon—from 180° C below zero to 100° C above—if it were not for the protection of its atmosphere. This protection is more than a simple matter of temperature. Short-wave radiation from the sun (x-rays, gamma rays, and ultraviolet rays) reach the moon unimpeded and would be deadly to an unprotected human there. The ionosphere begins to absorb these short waves and, between about 10 and 30 miles from the earth, a layer of ozone absorbs most of the remaining ultraviolet radiation. These short waves themselves created the ozone layer which prevents them from reaching the earth. Ultraviolet radiations separate the two oxygen atoms in some oxygen molecules, and these individual atoms then recombine with other oxygen molecules to form the three-oxygen-molecule of ozone. Because the energy of ultraviolet radiation is utilized in the constant creation, destruction, and recreation of ozone, little of it penetrates to the lower atmospheric layers. One reason that life is thought to have originated in the ocean is that before the development of the ozone layer it was only under water that life was sufficiently sheltered from ultraviolet radiation, yet the requisite oxygen for the ozone layer could not have come into being without the photosynthesis of living green plants.

In spite of its importance, ozone is one of the least abundant of all atmospheric gases. Although new ozone is constantly being produced, the highly-reactive ozone below the ozone layer is constantly being depleted by reactions with the elements and compounds which are much more abundant closer to the earth's surface. Increases in ozone during the day are therefore balanced by decreases during the night. If the ozone content of the upper atmosphere were much lower, life on the earth would be exposed to the destructive energy of ultraviolet radiation; if the ozone content of the lower atmosphere were much higher, life would be exposed to the toxic action of the highly-reactive ozone itself.

While nitrogen and oxygen total about 99 percent of the dry atmosphere's volume, the remaining one percent is much more diversified. Most of this (about 0.9 percent) is argon, a chemically inert gas. There are much smaller amounts of the other inert gases—krypton, zenon, helium, and radon. Although under certain circumstances a molecule of an inert gas will react with a molecule of oxygen, the reactivity of these gases is so low that they play almost no part in the chemical combinations, dissociations, and recombinations that are constantly occurring in the atmosphere. Radon, although chemically *inactive,* is radio*active.* It is therefore constantly decaying, but is replenished by the slow decay of uranium in the earth's crust. Radon is one stage in the gradual transformation of the unstable (and therefore radioactive) uranium 238 to the stable (and therefore nonradioactive) lead 206 and, as such, represents part of the natural background radiation which predates the nuclear age. Methane, also present in the atmosphere in small amounts, is produced by bacteria which decompose organic carbon or carbon dioxide and is the simplest of the hydrocarbons (compounds containing only hydrogen and carbon molecules). Nitrous oxide, one of the many possible combinations of nitrogen and oxygen, is present in a similarly small amount, as is hydrogen.

But, more important than any of these constituents of the last one percent of the atmosphere, both in atmospheric dynamics and in the interaction of the atmosphere with living things are carbon dioxide and water vapor. Carbon dioxide comprises about 0.03 percent by volume of the dry atmosphere and water vapor is concentrated mostly in the lowest layer of the atmosphere—the troposphere. At any one time and place, the air in the troposphere may be from zero to five percent water vapor.

The origin of the earth's water goes back to the days when the earth's crust was forming and this planet was blanketed by a thick layer of clouds (as Venus is now). The cloud water did not precipitate or, if it did, it immediately evaporated because of the earth's intense heat. As the earth cooled, rain fell and all the pockets in the earth's crust were filled with the precious liquid so essential to life. But not all of the water remained on the surface; at any one time, billions of tons of it are now present in the atmosphere as invisible water vapor or as clouds.

As for carbon dioxide, the atmosphere was once far richer in that gas than it is today. The temperature of the earth's surface was therefore warmer and plant growth was fantastically luxuriant—so much so that it outgrew the ability of animals and organisms of decay to consume it and return its carbon to the atmosphere. The carbon dioxide content of the atmosphere was thereby reduced, the surface temperature dropped, and carbon reached a state of equilibrium. Excess carbon in the dead plants ultimately became coal, oil, and natural gas.

Even the purest, most natural air in the days before man began to change the earth's surface and burn fossil fuels contained solid particles as well as gases. Volcanic eruptions sent great streams of particles high into the atmosphere and, closer to the surface, particles were produced by wind erosion of the land, forest fires, seasonal plant deposits, and salt tossed up by the sea.

This is the natural gaseous and particulate composition of the atmosphere. Also important among its life-sustaining properties are its density and pressure. Far out in interplanetary space, beyond the earth's atmosphere, the gas may have a density as low as about one atom per cubic centimeter. Dropping through the atmosphere toward the earth, we would find the density of the atmosphere increasing more and more rapidly until the air we breathe contains billions of molecules in one cubic centimeter. The atmosphere's pressure, of course, also increases as the earth is approached, until it is a thousand times greater at sea level than it is at the top of the stratosphere. The temperature of a gas is affected by both density and pressure; under normal atmospheric conditions, temperature increases as pressure increases (unless it is heated or cooled in other ways.)

It is in the lower reaches of the troposphere, below the highest peaks of the earth's crust, that the world of life—the biosphere—begins. Only in this air, within a relatively narrow range of temperature, pressure, and humidity, can life be sustained and reproduce itself. Here the atmosphere's composition and characteristics are completely compatible with the earth's myriad species. This is by no means coincidental: Life was suited to this atmosphere; it evolved in this environment, and played a large part in its production.

Energy of the Atmosphere

Visible light and near infrared radiation, the forms in which most solar energy reaches the earth, must be changed into other forms of energy in order to maintain the composition of the atmosphere, keep it in motion, power the earth's weather and climate, and allow life and breath. All of these necessary energy conversions and transfers are accomplished through the interaction of radiation and matter.

The sun's light energy that is absorbed by the earth's surface—land, water, and living things—is returned to the atmosphere as heat. This heat may then undergo a second conversion when it becomes *kinetic energy of motion.* When a mass of air is raised by warming from near the earth's surface, it has *energy of position.* Like a rock at the top of a slope, the mass of air has *potential* energy; both are subject to gravitational force. When the rock rolls down the hill, its potential energy is released. The potential energy of air may be released through the mediation of a temperature change, making it possible for the air to flow downward as shown on the following page. However, the most important role in mediating the release of the energy in the atmosphere is played by moisture.

Water. Hydrogen is the most abundant element in the universe, but very little free hydrogen is present in the atmosphere. However, there are more known compounds of hydrogen than there are of any other element. The most common and the one of greatest significance to life is water, which covers 72 percent of the earth's surface and makes up more than 90 percent of the human body. A water shortage in absolute terms is impos-

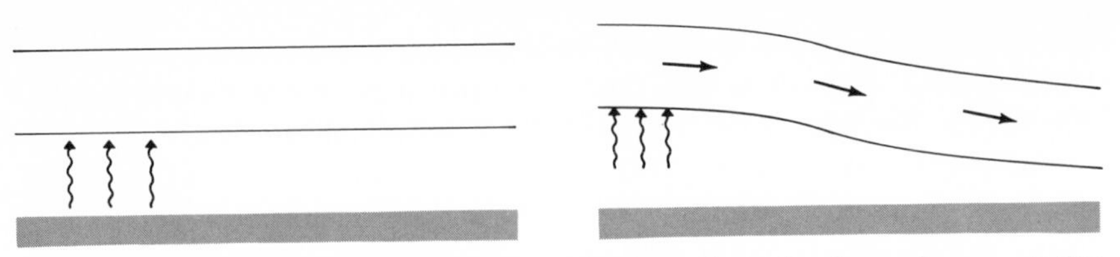

A simple air-motion induced by heating at the surface. A column of warm air tips the layer above and it begins to flow downward.

sible: The earth's water supply remains the same. But where the water is and in what form it occurs at any given time are of immense importance to living things. The water supply of a region can change dramatically with changes in the intimate and intricate interactions of evaporation and precipitation, of soil water and runoff, of icecap and ocean. In all these interactions, the atmosphere plays a part.

By far the greatest portion of the earth's water supply—97 percent—is in the oceans. The total amount of fresh water is difficult to measure because most of it is in polar icecaps, glaciers, and ground water. The remainder is divided between surface waters and the atmosphere in a respective ratio of about 100:3 units. But the small amount of water in the atmosphere circulates very rapidly from precipitation to evaporation and vice versa, turning over about once in every eight to eleven days—scarcely the tick of a clock in comparison with the slow movement of water through the oceans and the ice caps which we measure in thousands of years.

One of water's remarkable properties is that, within the range of temperatures and pressures common to the earth and its atmosphere, it can exist as liquid, solid (ice), or gas (water vapor) and when it changes from one of these states to another it takes up or gives off a tremendous amount of energy. In changing from liquid to vapor or from ice to liquid, water takes up energy and contains it in latent form; in changing from liquid water to ice or from water vapor to liquid water, it releases this energy. We use this property of water so often that we take it for granted. As perspiration is evaporated from our bodies, energy is withdrawn from our skin and the air surrounding it and our bodies are cooled. When a bottle of beer and a chunk of ice are placed in a picnic cooler, the ice melts, withdrawing heat from the bottle of beer and cooling it. In a steam engine, heat from burning fuel is transferred to water and, when the water is heated to the boiling point, its state is changed to steam (water vapor); the steam then contains latent heat or thermal energy which can be transformed to mechanical energy.

The atmosphere, too, makes use of latent heat, and the amounts of energy that are transferred in atmospheric processes make even the most powerful steam engine look weak by comparison. Changing only one gram of liquid water to water vapor by evaporation withdraws more than 500 calories of energy from the air around it; the rate at which water changes from one state to another in the atmosphere amounts to about 19 million tons per second. Water vapor may be carried hundreds of miles, bearing latent heat with it. When it changes its state once more and is condensed to cloud droplets, the latent heat is released, providing a tremendous input of energy for driving another aspect of the atmosphere's circulation—perhaps a storm—or for warming cold air in this other part of the world.

In some parts of the earth, precipitation exceeds evaporation. This is true in the extreme northern and southern latitudes (from 40 degrees poleward) where little heat is available for evaporation. It is also true in the proximity of the equator where, with more heat available, the rate of evaporation is much higher, but is nevertheless exceeded by an even higher rate of precipitation, including frequent cyclonic storms with their accompanying heavy rains. In the mid-latitudes, the reverse is true: More water is evaporated than returns to earth in rain and snow. Dividing the earth's surface in another way, there is more evaporation over the oceans and more precipitation over the continents.

There is also an inequality between the hemispheres, with more evaporation in the Southern Hemisphere and more precipitation in the Northern Hemisphere. Run-off returns surface water via the river systems and oceans from areas of excess precipitation to areas of excess evaporation. Yet this leaves a gap in the cycle. Obviously, part of the earth would become rapidly drier and the other part rapidly wetter were it not that the circulation of the atmosphere constantly closes the cycle by carrying moisture from the latitude bands with higher evaporation rates to those with higher precipitation rates. This includes transfer across the equator, keeping the hemispheres in balance, too, in spite of the fact that the basic wind patterns are intra- rather than interhemispheric. The much larger continental land masses in the Northern Hemisphere radiate more heat than does the watery Southern Hemisphere. Warmed air then rises on such a massive scale that colder air is pulled in at the surface from great distances—eventually from below the equator. Much of this interhemispheric air motion which transports moisture is probably accomplished by monsoons in the Indian Ocean.

The restless atmosphere seldom permits the water that has evaporated in any region to precipitate in that same region, but carries it away to some other place while it brings in a new supply of moisture that has evaporated upwind. Once a molecule of water vapor is high enough in the atmosphere, it may travel more than a thousand miles before it is precipitated.

Evaporation, as the word has been used here, is really two processes: the transformation of liquid water from both land and water surfaces into water vapor, and *transpiration* or the giving off of water vapor by plants. *Evapotranspiration* more properly expresses both processes. The evapotranspiration rate of any area depends upon its soil water content and type of plant cover as well as upon the amount and depth of its surface water, the temperature, the wind, and the vapor pressure of its atmosphere.

Precipitation tends to be higher when the atmosphere is unstable than when it is stable and higher when the air is cooling than when it is warming. How rapidly water vapor condenses into droplets and forms clouds is also affected by the number and composition of particles in the air, since certain kinds of particles can serve as condensation nuclei, initiating the formation of a cloud droplet. Whether—or when—a cloud precipitates depends upon the coalescence of many droplets forming a raindrop that becomes heavy enough to fall. It takes billions of water vapor molecules to form a single droplet and millions of cloud droplets to form a raindrop. When the rain finally comes down, it changes the composition of the air as it passes through it, washing out particles and, less efficiently, gases. (The efficiency of particle washout is far from complete; especially in the case of submicroscopic particles, many escape the drops of rain.)

It is only when we are subject to the destructive energy of a hurricane or some lesser, but still intense, storm that we get a hint of the immense kinetic energy of the atmosphere. It is only when we feel the heat of a midsummer day that we sense the immense heat energy of the atmosphere. We seldom realize that gigantic energy processes are taking place unceasingly in the atmosphere. Yet, in spite of their immensity, these energy processes are delicately balanced and a small change in one part of the system may have enormous repercussions. For example, it has been estimated that a temperature change of only *one-half of a degree centigrade* over the entire North Pacific can cause an increase or a decrease of billions of tons of annual evaporation from the surface[4].

Air in Motion. The atmosphere's energy of motion or kinetic energy, in spite of its immensity, represents a mere one percent of that part of the sun's energy that is intercepted by the atmosphere.

The general circulation—the great sweeping motion on a global scale—is what helps to make the earth livable. If heat were not moved from that part of the atmosphere receiving more of the sun's energy to that part receiving less, we would have a climate of much harsher contrasts—hotter in the hemisphere experiencing summer and colder in the one experiencing winter. Within each hemisphere, too, the north-south temperature differences would be much greater. The very existence of the temperature inequalities and the pressure inequalities that accompany them sets in motion the winds which moderate these differences. Currents of air in the polar and tropical regions tend to move near the earth's surface from colder regions toward warmer ones—rising as they are warmed and expand, returning to the colder regions at higher altitudes, and then sinking and again flowing near the surface toward the warmer regions. Between these polar and tropical currents, in the mid-latitudes, an indirectly-driven circulation wheel moves in the opposite direction.

The earth's major wind patterns, set in motion by these temperature differences, are strongly affected by the earth's motion as it rotates from west to east. This deflects the paths of winds to the right in the Northern Hemisphere and to the left in the Southern Hemisphere.

Many other temperature inequalities also initiate air movements. Land becomes hotter during the day and cools more rapidly during the night than

The atmosphere is constantly in motion, although not always in such violent motion as depicted here in a photograph of Hurricane "Gladys" taken from Apollo 7 on October 17, 1968. Storms can be destructive, but they also play an important role in transporting water vapor from ocean to land and from the Southern Hemisphere to the Northern Hemisphere as well as in transporting heat from the warmer tropics to the cooler latitudes near the poles.

a large body of water does, creating landward breezes in the daytime and seaward breezes at night, with the air aloft flowing in the opposite direction. The oceans store heat during the day which is given off at night, they store heat during the spring and summer which is given off during fall and winter. Water is particularly effective in moderating temperature inequalities in this way because of its high capacity for absorbing heat. It takes more energy to raise the temperature of a quart of water one degree than it does to produce the same temperature increase in almost any other substance. This high specific heat of water is also responsible for *initiating* some temperature inequalities; it is one of the reasons, for example, why water both heats and cools more slowly than land.

The heat of the ocean is redistributed by ocean currents and this ocean circulation, itself largely wind-driven, is at once a source of temperature inequality in the atmosphere above the ocean and one of the ways in which temperature differences between widely separated areas are moderated.

There are temperature inequalities between a mass of cold air moving into a warmer region and the air into which it is moving. Where the two masses of air meet, there is a *front* of strong temperature contrast where slow mixing of the two masses is taking place. (A similar front is found when a warm air mass—like the one that led to Episode 104—moves toward a colder region.) Temperature inequalities also occur between one land area and another because some varieties of plant cover retain heat better than others. And there are temperature inequalities between mountains (especially if they are high and snow-covered) and lowlands, with cold air flowing down the mountainsides toward the warmer lowlands.

Where the earth is warmer than the air just above it, air movement is initiated. A rising column of warm air striking a layer of cooler air above it can tip this layer and start a downward flow of the cooler air. Warm moist air is even lighter than warm dry air, so the humidity of the air affects its vertical motion. A tiny, upward puff of air may immediately subside, or a parcel of warm air may be buoyed up by colder, heavier air flowing in below and attain considerable height before it cools and sinks again. Air rises least and returns to the earth most quickly if the atmosphere is stable—that is, if it is not losing heat rapidly with altitude. As a parcel of air rises, the rate of temperature decrease resulting from a decrease in pressure and density is called the adiabatic lapse rate. (*Adiabatic* means without gain or loss of heat.) The temperature of a parcel of air will decrease adiabatically about 10° C for every kilometer it rises. (This is roughly equivalent to a decrease of 18° F as it rises two-thirds of a mile.) But there are usually gains and losses of heat from other factors besides the changes in pressure and density with altitude. If the decrease of temperature with altitude is smaller than the adiabatic lapse rate because the sun is warming the air, for example, then the atmosphere is stable and vertical air movement scarcely gets off the ground. If the decrease of temperature with altitude is greater than the adiabatic lapse rate because a mass of cold air is moving into the area, for example, then the atmosphere is unstable and pockets of air rise and fall rapidly.

Of course, a "parcel" of air is not neatly wrapped and tied like a postal parcel and is not moving smoothly in one direction. Air that begins to rise may be tossed and tumbled about by friction with surface objects and irregularities. It may mix with other parcels of air by exchanging particles and molecules of water vapor and other gases. It may be swept into a horizontal wind or become part of an *eddy*—a current moving at variance with the dominant wind direction. The mechanical turbulence induced by friction as air moves across an uneven surface and the thermal turbulence induced by the heating and cooling of adjacent small parcels or great columns of air affect the atmosphere's motion far above the earth.

Only at a height of a mile or so above the surface is the atmosphere

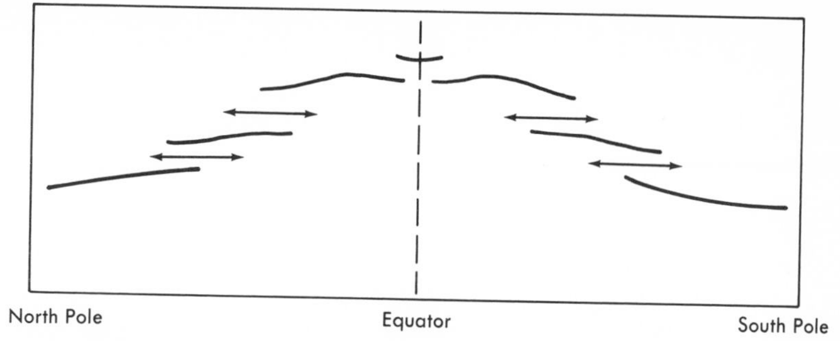

Air is exchanged between the troposphere and the stratosphere through "windows" in the troposphere where its altitude changes. The tropopause is indicated by the heavy lines; the exchange by the double arrows.

relatively free of the effects generated by these multitudinous surface variations. At heights here and above, atmospheric motion is much simpler. Wind speed increases with altitude and, in the jet streams at the top of the troposphere, reaches more than 150 miles per hour. These high, fast-moving winds exert a strong influence on the weather below. Above them is the tropopause, the transition between the troposphere and the stratosphere, where the temperature which has been dropping with altitude reaches a minimum before beginning to rise again in the stratosphere under the influence of the ozone layer which is heated by the trapped ultraviolet radiation. The stratosphere has its own wind and weather patterns with occasional enormous and rapid temperature changes, but how much they affect tropospheric weather and climate is not well understood.

Across the tropopause, exchange goes on between the upper and the lower atmosphere through "windows" in both northern and southern latitudes, as shown in the figure above. Exchange also occurs during the powerful suction of a cyclonic storm which pulls stratospheric air down into its vortex while, around the edges of the storm, tropospheric air rises into the stratosphere. Thunderclouds can also produce remarkably strong updrafts which can move with speeds of more than 60 miles per hour and can break through the tropopause into the stratosphere.

Atmosphere and Life

This is the air we breathe—always changing, yet always remarkably the same. Our knowledge of its numerous balancing acts is far from complete, but we do know that without its constant interactions with living things, life would be impossible.

Green plants obviously occupy a crucial position in the biological cycle which moves from life to growth to death to decay to life again. If this single process is resolved into the interlocking cycles of energy, water, oxygen, carbon, nitrogen, and a variety of nutrients, green plants lose none of their importance. From the phytoplankton that live only a few days and are so

Green plants stand at a crucial intersection of the energy, oxygen, carbon, water, and nutrient cycles.

small that they are visible only as the green color of the ocean water, to the fields of corn planted and harvested seasonally by man, to the giant redwoods that may live more than three thousand years, green plants stand at a critical intersection of these cycles and the atmosphere plays an important part in almost all of them.

Energy. Only about a thousandth of the solar energy reaching the earth is captured by green plants and enters biological processes through photosynthesis. Some of this solar energy is used in the plants' own life processes which culminate in the release of heat as well as oxygen and water vapor, thus removing energy from the biological cycle and returning it to the atmosphere. This is the most rapid turn of the energy wheel. What energy is left is stored in the cells of plant tissue as the potential energy of the chemical bonds that hold the organic molecules of the cell together and in the attractive and repulsive forces of the ions in the cell—atoms and molecules that are positively or negatively charged. Directly through plant food, indirectly through animal food, plant tissue is thus the source of all the energy required by man for the internal physiological processes of growth and reproduction and all the external activities of human life. The life process of man and other animals, like that of plants, releases heat, removing energy from the biological cycle and returning it to the atmosphere in a somewhat slower turn of the energy wheel.

Energy from photosynthesis moves through the food chains of decay as well, with dead plants, the bodily waste of live animals, and the bodies of dead animals being decomposed by insects and microorganisms, making inorganic nutrients available for a new cycle of growth. Some of this material is not totally decomposed into simple inorganic nutrients, but remains in the form of more complex carbon compounds and retains the potential energy of their chemical bonds. Stored in bottom sediments of bodies of water over millions of years and under certain conditions, some of this potential energy may become the highly concentrated stored energy of fossil fuel.

Since every conversion of energy is accompanied by some loss of energy in the form of heat, it is not only in the transpiration of plants, but in each step of the energy cycle, that waste heat is given off and is returned to the atmosphere (sometimes via soil or water, sometimes directly) to warm the air, evaporate water, or move air currents, and eventually to become part of the heat that the earth radiates back into space. Meanwhile, the biosphere's energy is constantly being renewed as the atmosphere continuously delivers a new supply of solar radiation to the green plants. The amount of energy going into prolonged storage is very small; the energy entering the biological cycle from photosynthesis is almost evenly balanced by the energy that is being returned to the atmosphere in the respiration of plants and animals, including the microorganisms of decay.

Oxygen. Because oxygen readily combines with almost every other element, it is present throughout the biosphere in a tremendous number of compounds, but living things need it in its free form. Ultimately, this free oxygen is derived from water: Part of the photosynthetic process is the splitting off of the oxygen atom from H_2O and its combination with another oxygen atom, thus forming free oxygen.

Within the human body and all other aerobic forms of life, living cells obtain energy for heating the body and for carrying on its processes by breaking down carbohydrates, fats, and proteins. This is done by a complex series of chemical reactions. These begin with a fermentation process which does not require oxygen and proceed to the much more efficient aerobic process somewhat analogous to fire releasing energy by breaking down fuel in the presence of oxygen. In this oxidation process, the free oxygen becomes bound once more to atoms of carbon in carbon dioxide and to atoms of hydrogen in water vapor. The carbon dioxide and water vapor return to the atmosphere in exhalation or, less directly, after passing through many chemical changes, as the end products of decomposition.

Oxygen from the atmosphere mixes with the waters of the lakes, rivers, and oceans; thus these waters contain free, dissolved oxygen as well as the oxygen that has combined with hydrogen to form water molecules, making possible the many forms of aquatic life. Oxygen from the atmosphere penetrates the soil, making possible the functioning of plant roots and the intense microbiology of the soil. And oxygen from the atmosphere constantly reacts with various minerals in the earth's crust to form mineral oxides.

The oxygen cycle itself is therefore a complex interaction of cycles, with oxygen existing alternately in free and combined states and moving through the land, the water, living things, and the atmosphere. By these many complicated and interacting pathways the oxygen content of the atmosphere is maintained and the life of oxygen-demanding animals like man is made possible.

Carbon. As essential to life as oxygen, carbon is much less abundant. (There is almost 700 times as much oxygen as there is carbon dioxide in the atmosphere.) The carbon cycle is therefore more sensitive to quantitative change (a doubling of carbon dioxide would mean only a negligible drop in oxygen). By far the largest part of the carbon that occurs naturally in the atmosphere is in the form of carbon dioxide added principally by the oxidation of methane emitted by decaying organic matter. In the biosphere, carbon atoms can attach themselves to one another in complex chains and rings and form an almost unending number of chemical compounds. Once again, these all begin with photosynthesis.

Green plants use carbon dioxide to manufacture carbohydrates—the simple sugars that combine carbon, hydrogen, and oxygen and that are the first of the many carbon compounds in living matter. As plants withdraw carbon dioxide, atmospheric concentrations vary daily near the ground and seasonally all the way to the stratosphere. An increase in carbon dioxide in the atmosphere can cause an increase in photosynthesis up to a limit which varies among species and is dependent upon light intensity and temperature.

The carbon cycle, like the oxygen cycle, is a series of wheels within wheels—all related, but moving at different speeds. The carbon dioxide that is drawn *into* a plant may be fixed within a matter of hours, but the turn of the wheel that carries that same carbon dioxide *from* the plant back into the atmosphere is much slower. Some carbon dioxide is released when dead plant material is broken down into humus; still slower is the liberation of carbon dioxide from the more complex carbon compounds in the soil, which can take decades in tropical soil and hundreds of years in a northern forest.

Carbon dioxide from the atmosphere is also dissolved in the surface layers of the ocean. It is more soluble at a lower temperature, so it is dissolved most rapidly in the Arctic. At the same time, dissolved carbon is being taken up by the atmosphere from the ocean, most rapidly in the tropics. But there is a more complex and much longer process of exchange between the surface layers of the ocean and the deeper layers: Mixing by deep ocean currents, deposits of mineralized carbon in bottom sediments, and upwelling from ocean bottoms toward the surface again takes at least a thousand years.

Finally there is the slow storage, transformation, and concentration of

What is happening in the atmosphere cannot be understood without understanding the constant interchange of heat, gases, and particles between the air and the oceans over almost three-quarters of the earth's surface.

dead plant material into coal, oil, and natural gas—a process that takes millions of years.

Nitrogen. Since nitrogen is even more abundant in the atmosphere than oxygen, the nitrogen cycle would appear to be the most impervious to change. But like the abundant radiation of the sun, the nitrogen of the atmosphere can enter the biological cycle only through a very narrow channel. Much of living material, including the human body, is made up of nitrogen-bearing molecules—protein, nucleic acids, enzymes, vitamins, and hormones. Yet most living things take in nitrogen from the air and release it again without using it. The needed nitrogen must be combined with other elements *before* it enters plants or the bodies of animals.

Intense heat in the atmosphere, as from lightning, can combine nitrogen with oxygen to form nitrogen oxides which can, in turn, be washed from the air and deposited in soil or water as nitrates and then drawn into plants through their root systems. But this produces very little of the nitrogen in the biological cycle. Nitrogen enters living things almost completely by way of certain microorganisms that are capable of combining nitrogen from the air with hydrogen[5]. This nitrogen fixation is carried out by some species of bacteria which live free in the soil and others which are associated with the roots of legumes or the leaves of some tropical plants. Once fixed in this manner, the nitrogen moves repeatedly through a plant-soil-plant cycle in organic compounds with carbon and hydrogen and then back into inorganic compounds with oxygen. In surface waters, a similar plant-water-sediment-plant cycle takes place. The nitrogen used by animals is returned to the soil or water by way of animal waste. In both soil and water, only a little free nitrogen is returned to the atmosphere by denitrifying bacteria which separate it from the oxygen in inorganic nitrogen compounds. Under natural circumstances, denitrification probably just balances the input of nitrogen being fixed by such nitrogen-fixing organisms as the bacteria that live symbiotically with legumes and with trees.

As with other elements, there is a geological nitrogen cycle: Inorganic nitrogen is removed from the biological cycle and stored in sediments, ultimately to become sedimentary rocks, while the weathering of rocks on the earth's crust as well as volcanic action provide the atmosphere with a new supply of nitrogen.

Sulfur. In addition to nitrogen, oxygen, carbon, and hydrogen, many other elements are essential to life in much smaller amounts. Sulfur is the only one of these that is volatile and that circulates naturally through the atmosphere. It is abundant in the earth's crust, having been deposited over a long period of time in sediments and incorporated into sedimentary rock. Atmospheric sulfur may appear in the form of sulfur dioxide, but, by the time it is brought down to the earth in rain, it is further oxidized to sulfates and is taken up through the roots of land and aquatic plants and becomes available to animals. Or sulfur may be washed out of the soil, carried to the ocean, and some of it returned to the atmosphere in salt spray. Anaerobic bacteria, living in the mud of swamps, streams, ponds, lakes, and coastal waters, reduce sulfates from decomposing matter to hydrogen sulfide—a gas which also returns the sulfur to the atmosphere. Once in the air, hydrogen sulfide is quickly oxidized to sulfates and the cycle turns once more.

Pause for Breath

What we already know about the atmosphere inspires us with awe at its complexity and the resilience which arises from this complexity with all its many possibilities for internal readjustments. But we are also becoming increasingly aware of the atmosphere's fragility and of its limits. If we are to understand this precious resource well enough to maintain its great life-giving potential for future generations, we have much to learn. In our use of the atmosphere for waste disposal, do we rely too much on its resilience? In our ignorance, are we stressing the web at some fragile—perhaps crucial

—intersection? Or interfering in one of the natural cycles so seriously as to unbalance it?

The search for knowledge will have to be both an interdisciplinary and an international effort, for the atmosphere defies fragmentation into specialties or confinement within national boundaries. And the questions for which we seek answers are of immediate concern, not just to scientists and governments, but to everyone who breathes. This is a divided world in many ways, but it is inescapably one in the world of the atmosphere.

[1] The *solar constant* is the amount of energy received at the outer edge or fringe of the earth's atmosphere, at right angles to the sun's rays and at the average distance from the sun (since the earth and its atmosphere are not always at the same distance from the sun). The value of the solar constant is two calories per square centimeter per minute. (A *calorie* is the amount of energy required to raise the temperature of one gram of water from 14.5 °C to 15.5 °C.)

[2] J. M. Mitchell, Jr., "Recent Secular Changes of Global Temperature," *Annals of the New York Academy of Sciences,* 1961, 95:235–50.

[3] A. I. Oparin, *The Origin of Life* (New York: The Macmillan Company, 1938).

[4] J. Namais, *Journal of Geophysical Research,* 1970, 75:565. Quoted by Helmut E. Landsberg, "Man-Made Climatic Changes," *Science,* December 1970, 170:1265–74.

[5] A recent study indicates that the leaves of plants also absorb significant quantities of ammonia (a compound of nitrogen and hydrogen) directly from the air. G. L. Hutchinson, R. J. Millington, and D. B. Peters, "Atmospheric Ammonia Absorption by Plant Leaves," *Science,* February 18, 1972, 175: 771–72.

The material in this chapter was drawn from a variety of sources, of which those listed below may be useful for additional reading. The book by Chandler is especially helpful for those with no background in physics and mathematics.

The Biosphere. Special issue of *Scientific American,* September 1970.

Chandler, T. J. *The Air Around Us.* Garden City, New York: The Natural History Press (in association with Aldus Books, Ltd., London), 1967.

Davis, Kenneth S. and John Arthur Day. *Water, The Mirror of Science.* Garden City, New York: Anchor Books, Doubleday & Co., Inc., 1961.

Flohn, Hermann. *Climate and Weather.* Translated from the German by B. V. de G. Walden. New York: McGraw-Hill Book Company, 1969.

Sellers, William D. *Physical Climatology.* Chicago: The University of Chicago Press, 1965.

The Burdened Atmosphere

Gaps in our knowledge of the natural atmosphere become chasms when we try to understand how man is changing it. To bridge these chasms we must have accurate, integrated measurements of moisture, temperature, pressure, wind, and clouds over the whole globe and throughout the various layers of the atmosphere. Data from over the oceans are scanty compared to data from over the continents, and we have more information from the Northern than from the Southern Hemisphere, but meteorology has made rapid progress in recent years and, with the introduction of weather satellites, another tremendous step forward has been taken.

A Global Atmospheric Research Program (GARP) has been planned by the World Meteorological Organization and a Joint Organizing Committee of the International Council of Scientific Unions to gather the data needed to improve our short-term weather predictions, and some of the research is under way[1]. Further investigations of long-term climatic change and how man is affecting it are also needed. We must learn more about the atmosphere as a heat engine—that is, how solar and terrestrial radiation affect the earth-atmosphere system; we need to measure the gaseous and particulate composition of the air around the world and record its changes over a period of years. We must be able to test theories of climatic change against real, measured climatic changes at least on a short-term basis before we can have confidence that these theories will help us predict the long-term effects of interfering in a massive way with the atmosphere's natural processes. We need to know how these changes affect all other aspects of

atmospheric composition and motion and how they interact with land, water, and living things.

Broad public support will be needed for both basic and applied science if we are to gain the much deeper knowledge we need of the only world we have. The mission of this scientific effort is not a single spectacular accomplishment, like harnessing the energy of the atom's nucleus for military purposes or putting a man on the moon. Rather it is the human mission of surviving on this planet—a purpose immediately relevant to every human being, but so broad in scope that it is easy to lose sight of the meaning of the whole in the variety and multiplicity of its parts.

Support will come more readily if scientists keep people informed of the current state of knowledge, including frank discussions of what is not known. This is essential for a more immediate reason. Decisions that affect the environment are made every day; they are determined by many other considerations besides scientific information, but, however partial our knowledge, these decisions should reflect the best available information. We readily recognize that it requires a decision to reduce or halt some form of pollution. What we often forget is that decisions are also made to *continue* or to *increase* pollution because we are not sure that such actions will be harmful. It would take a revolution in our thinking to halt some particular agricultural or industrial activity which will affect the environment because we are not sure its effects will be harmless. Yet because even small changes in the earth-atmosphere system can have mighty effects on this uniquely life-giving planet, we must recognize that our margin for error is also small.

As we examine a series of human interventions in the atmosphere, it is well to keep in mind that our information can never be complete. Although we are a long way from the understanding we need, we continue to intrude at an accelerating rate into the atmosphere's natural composition and characteristics. Some of this interference is an unintended side effect of other human activities; other interventions have been proposed in the name of weather and climate modifications that—it is hoped—would prove beneficial. So it is not enough to seek additional understanding; it is necessary to test what we are doing (and what we propose to do) against what we already know of the possible consequences of our actions, to find ways of balancing both known and possible hazards against desired benefits, and to do this again and again as our growing knowledge makes a more realistic balance possible.

Intentional Modifications of Weather and Climate

It is typical of our confidence in technology that we are already trying to do something about the weather, not just by dissipating fog at an airport or inducing rain in some small area during the growing season, but by proposing large-scale modifications in weather patterns as well. Weather modification has been the purpose of much recent research. In fact, GARP originated in the United Nations following a speech by President John F. Kennedy in 1961 proposing "further cooperative efforts between all nations in weather prediction and eventually in weather control."[2] Specific proposals thus far have involved changing the heat balance of the earth in order to warm regions that are now too cold for agriculture, increasing precipitation over large areas that are now too dry, and reducing the destructive impact of hurricanes.

Recently two meteorologists independently worked out models of the global climate and reached the same conclusion: *Our present climate is delicately balanced*[3]. On the one hand, a decrease in the solar constant of 1.6 percent (according to M. I. Budyko, a Soviet scientist) or 2–5 percent (according to W. D. Sellers of the University of Arizona) would be enough to initiate a new ice age. The same effect might be achieved by a corresponding increase in the *albedo* (the reflectivity) of the earth. Similarly, increases in the solar constant or decreases in the earth's albedo would not have to be very great to melt the polar icecaps.

No one wants to induce a new ice age, but it is less obvious that melting the glaciers would be undesirable. Yet among the possible consequences of the latter would be smaller temperature differences between latitudes, which would affect large-scale atmospheric circulation and consequently the transport of moisture in as yet unpredictable ways. Neither does anyone know the effect of the resulting changes in the flow of energy and water through the various ecosystems[4]. As the ice on land melted and the water ran down to the sea, the level of the oceans would gradually rise, slowly inundating seacoast cities and creating catastrophic problems for people. Again, what the effect of this change might be on the ecosystems of estuaries and continental shelves is unclear and would depend in part on how rapidly it occurred. Since the world climate in past ages has varied widely, the hypothesis that the present climate is unstable suggests that human activities could speed up a natural trend in either the direction of a new ice age or of glacial melting.

One ice-age theory suggests that the very melting of the Arctic icecap could produce a new ice age[5], contradictory as that may appear. Maurice Ewing and William Donn point out that if the icecap melted, the amount of moisture in the atmosphere would be greatly increased. This moisture-laden air moving over the colder land masses in winter would precipitate snow which, in the short Arctic summers, would not melt completely, causing the snow blanket to grow thicker each winter. Meanwhile, the inflow of warm water from the Atlantic Ocean would prevent the polar icecap from reforming in the Arctic Ocean and ice would continue to accumulate on the continents until vast glaciers were formed. These glaciers would gradually move down the continents, as in the last glacial epoch when the ice reached into northern Illinois (see map on the following page). As more of the earth's supply of water became stored in the glaciers, their growth would be paralleled by a lowering of the level of the oceans. Ultimately, the Atlantic would fall about 300 feet, with the result that there would no longer be a sufficient flow of warm water over the Norway-Greenland sill to keep the Arctic Ocean from freezing again. The glaciation process would then at last begin to reverse itself: The snow would melt faster than it formed, the icecap would gradually accumulate, and the earth would slowly emerge from the ice age. The whole process might take millions of years.

A change in climate, even a much less massive change than a new ice age, can unbalance other parts of an ecosystem. Sunlight, temperature, and rainfall are intimately related to the life patterns of thousands of living things. When a climate changes slowly, these organisms have many generations to adapt to shorter or longer winters, to greater or lesser amounts of rain, to more or less sunshine. But when a climate changes suddenly, there is inadequate time to allow adaptation by natural selection. The disruptive effects of man-induced changes in climate are difficult to predict, but could include the elimination of species in an area where they formerly flourished and a consequent reduction in the diversity and the stability of a particular ecosystem.

With all our vaunted independence of the weather, climatic changes could require difficult and expensive readjustments in human patterns of life as well. Anyone who has lived through a snowstorm in a city which seldom has snow and is therefore unprepared to cope with it, can readily imagine some of the problems that might follow if St. Louis began to experience Minneapolis winters, while a need for space heating on a large

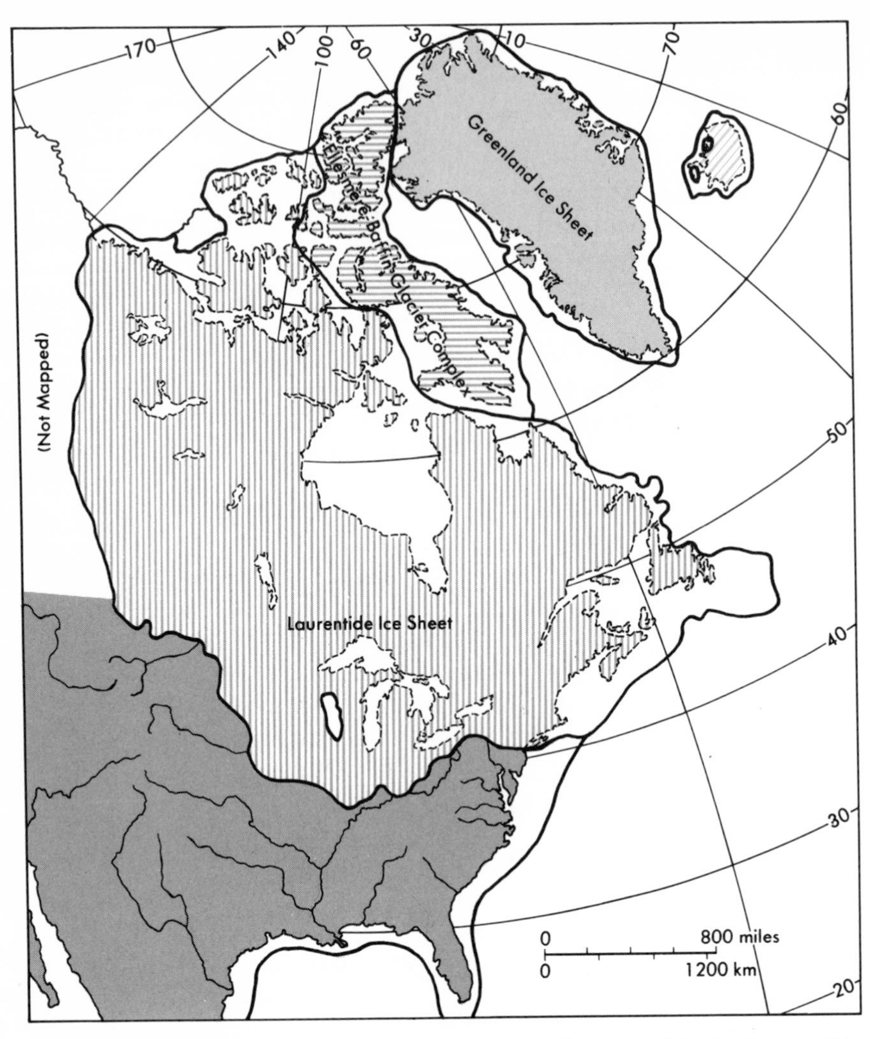

We are playing a dangerous game by interfering with the global heat balance in this interglacial age. A decrease in the temperature of the far North might initiate a new ice age, with glaciers creeping South as shown in this map which outlines the farthest extent of glaciers in the last ice age. With more of the earth's water frozen and less of it in the oceans, the shoreline of the North American continent during the last ice age may have been as indicated here by the heavy line.

scale in Los Angeles would render many of that city's buildings obsolete, change its energy needs, and worsen its air pollution problems. Agricultural patterns are even more dependent upon weather that fluctuates within expected ranges of temperature and rainfall. More than one crop could well be lost before planting and harvesting were adapted to a new weather pattern. In those parts of the world where people live on the edge of starvation, this could spell famine.

A number of rainmaking experiments have advanced cloud-seeding technology considerably[6], although success in increasing precipitation has been achieved under only limited atmospheric and topographical conditions. (Under certain conditions, seeding can actually *decrease* rainfall.) For experiments in weather modification, the environment is the laboratory. The danger of an experiment at the wrong time and place was demonstrated when a rainmaking experiment in the summer of 1972 was carried out near Rapid City, South Dakota, while a storm was gathering. Experimenters claimed that the cloud-seeding had no effect on the subsequent 14-inch rain and the flood that followed, but this is impossible to prove—just as it is impossible to prove that the experiment augmented the rain. It is clear, however, that there was inadequate understanding of the weather conditions and poor judgment in seeding clouds which *could* contribute to the severity of the rainfall[7].

Weather modification as a weapon of war has been discussed as something that might be part of a war of the future[8]. It now appears that the United States has been carrying out both experiments and operations in rainmaking in Indochina[9]. The nature, extent, and results of these operations are classified, but the purpose is said to be making transport and antiaircraft defense more difficult.

One current experiment, Project Skywater in the upper Colorado River Basin, is attempting to increase the snowfall in the San Juan mountains in the hope of adding, through runoff, to the water in the Colorado River system. If the pilot project is successful, questions about possible side effects, as well as cost, must be resolved; studies are now underway to try to determine whether or not a full-scale project would increase the number and severity of avalanches, reduce precipitation downwind, or produce other undesirable climatic changes or effects on plants and animals.

Hermann Flohn, in *Climate and Weather,* while cautioning against irresponsible interference with water and heat balances, describes a scheme proposed by the Swedish meteorologist T. Bergeron as being based on sound climatological principles[10]. The area concerned is the Sudan Belt, about 600 miles wide and more than 3000 miles long, which stretches across the broad part of the African continent from the Atlantic Ocean to the mountains of Ethiopia. The entire flow of the Niger and Nile rivers and much of the Congo and Lake Chad would be used for irrigation according to Bergeron's plan, increasing evaporation during the summer in an area where air is generally moving upward. This would stimulate cloud formation and subsequent precipitation, thus using the same water over and over again to enhance agricultural productivity. But sound climatology is not necessarily sound ecology: Less massive interventions into the water cycle have produced unfortunate side effects[11].

Project Stormfury is one experimental weather modification program with a goal—abating the fury of hurricanes—that appears to be an unadulterated good. Clearly, it is desirable to save lives and prevent property damage. However, if the goal becomes attainable, where and when the abatement techniques are used will have to be carefully selected[12] to

One of the goals of weather modification is to abate the fury of hurricanes and lessen such destruction as this, caused by Hurricane Camille in Biloxi, Mississippi, in August 1969. However, it must be remembered that storms also play an important part in the circulation of heat and water.

assure that the land is not deprived of needed moisture; the Atlantic seaboard is heavily dependent upon storms for rain. There may also be a danger that in interfering with storms we will also interfere with the transportation of heat from the equator toward the poles and of moisture from the Southern to the Northern Hemisphere.

There is general agreement about the need for more information before intervening massively in climate and weather, but we are in danger of solving the technical problems of *how* to do it before we are able to solve the scientific and social questions of *whether* to do it, as has happened so often in the fields of industrial, agricultural, and military technology in the recent past. If an oceanic storm can be modified before it strikes land, will this deprive that coastal area of needed water while protecting it from physical damage? If so, how can the protection be balanced against the water loss? If an increase in rain can be achieved in a dry area, will it deprive some other county, state, or nation of needed moisture? If so, who is to decide over which land the rain should fall?

The effect of a given change in rainfall on an area's ecosystem is the result of its influence on the complex interrelationships among many organisms. Human interrelationships may also be affected. The added rain that improves the farmer's crop may ruin the resort-owner's tourist business. Some thorny legal problems have already arisen from conflicting interests in weather modification[13].

The more complex the weather modification plans become, the more difficult are the scientific questions and the harder they are to answer on an experimental basis. The social, economic, and political questions are correspondingly bigger and more complicated. The social questions become still more difficult to resolve when the scientific answers are equivocal. Is the known benefit worth the possible—but uncertain—risk?

Meanwhile, we continue to introduce changes in the earth-atmosphere system inadvertently as we pursue other goals[14]. These side effects may be both unanticipated and unwanted.

Interfering with the Global Heat Balance

Man's conversion of stored energy to heat began with the discovery of fire, but remained miniscule for many thousands of years. The burning of wood, coal, oil, gas, and nuclear fuel still produces an amount of heat so small in comparison with the solar energy converted to heat in natural processes that its effect on global climate is usually assumed to be negligible. Whatever the effect may be, it cannot be measured or even estimated with any known technique, but the steep increase in man's energy consumption since 1940 suggests that this may not continue to be true. The United States and the U.S.S.R. now account for 40 percent of the world's annual energy consumption. If the pattern established by these two countries were followed by the rest of the world, the heat man adds to the atmosphere would increase a hundredfold in about a century and would then represent a perceptible fraction of the heat received from the sun.

This thermal pollution is much more intractable than the gases and particles we usually think of as air pollution. No pollution control technology reduces it. Although we speak of energy "consumption," it is impossible to truly "consume" energy. In the first transformation from stored energy to mechanical or electrical energy, some waste heat is given off, and no matter how we use the energy, eventually it is *all* dissipated into the atmosphere. Its effect on global climate may depend somewhat on changing fuel use. Fossil fuel is accompanied by carbon dioxide and particulates which also may affect the climate, while nuclear fuel brings a rise in radioactive pollution and more waste heat per unit of electricity generated.

Man is already converting stored energy to heat on a very large scale within particular land areas and in particular latitude belts, so the temperature differences which now initiate large-scale atmospheric motion between land and sea and between North and South may become greater in some places, with a possible increase in storms. Temperature differences may be reduced in other places, decreasing atmospheric motion and consequently changing the distribution of energy and water vapor. If regional changes are big enough, they may add up in unexpected ways to changes in the circulation of heat and water, with profound consequences for life. A warming of the lower atmospheric layers would be accompanied by a cooling of the upper layers; this would make the atmosphere more unstable and might increase high-level cloudiness, with further unpredictable consequences.

Our predictive tools are limited, so it is hard to see where we are headed as we increase the waste heat from manmade sources. Mathematical models of atmospheric processes can offer only a tentative estimate at best. One such model suggests that in a little more than a hundred years temperatures might rise as much as 27°F near the North Pole, but exhibit very little change near the equator[15]. Another model suggests that if we continue to increase our output of waste heat at the present rate, global temperatures could rise in roughly 300 years by an average of 27°F, ranging from an increase of about 20° near the equator to about 48° at the North Pole. This "should be enough to eliminate all permanent ice fields, leaving only a few high mountain glaciers on Antarctica and perhaps Greenland."[16]

Will we really continue to increase our thermal pollution of the atmosphere at the current rate? That depends on many economic, social, and technical factors. But whatever the rate of increase and whatever the limitations of our present ability to predict its climatic effects, indications are that there is an upper limit to the amount of waste heat we can dissipate into the atmosphere without producing major effects on the global climate.

We may be affecting global heat balance in another way—by rapidly releasing the carbon that has been stored in coal, oil, and gas over millions of years. The carbon dioxide content of the air has been increasing quite

The amount of surface water locked in glaciers and polar icecaps is regulated by global temperature and, in turn, is one regulator of the global climate. Man's release of pollutants (including waste heat) may be sufficient in the foreseeable future to change this present balance either by melting the icecaps or by increasing their extent, thus setting off a complex series of changes in the global climate. This oblique air photograph shows the margin of Vestfonna sheet in Svalbard, Norway.

steadily with our increase in the burning of these fossil fuels, and we are therefore steadily reinforcing the ability of the natural atmosphere to absorb and re-radiate more of the infrared radiation given off by the earth[17]. It now appears that half the carbon dioxide added by combustion remains in the atmosphere, while the other half is absorbed by the upper layers of the oceans and by green plants. But will this continue to be so? How much carbon dioxide goes into each of the two reservoirs and what are their limitations? How soon will the carbon newly stored in living organic material be returned to the atmosphere? How long will it take for the carbon dioxide in the upper layers of the oceans to reach the lower layers and ultimately become stored as inorganic carbonate and bicarbonate, thus permitting the absorption of more of the excess carbon dioxide in the oceans' surface layers? There are as yet no firm answers to these questions.

Higher temperatures, whether caused by waste heat or carbon dioxide or both, could reduce the solubility of carbon dioxide in ocean water and cause the oceans to release more of that gas into the atmosphere. Increased temperatures would also increase the rate of evaporation and the reflectivity of the earth would be reduced by melting snow and ice. All of these effects would tend to further increase the temperature.

One estimate of the effect of increased carbon dioxide is that an increase in global temperature of 0.9 °F will occur by the year 2000 and perhaps of 3.6 °F if the carbon dioxide content of the atmosphere is doubled[18], which might occur in some 400 years. Another view is that, after an initial rise, the rate of temperature increase would be proportionally less and less for additional carbon dioxide and would finally level off[19].

Were man's interference in the atmosphere limited to one pollutant at a time, its effects would not be so difficult to trace. But accompanying the carbon dioxide is an increasing volume of particles from industrial and agricultural activities which may have a cooling effect and thereby offset the effect of carbon dioxide. This might account for the fact that world temperature is now dropping, although the carbon dioxide content of the atmosphere continues to rise.

At the time of atmospheric nuclear testing, when streams of particles were thrown into the stratosphere, their radioactivity made it possible to follow them from one atmospheric layer to another. The average time a particle remains in the various layers was observed to be as follows:

Atmospheric Layer	*Average Lifetime of Particles*
Lower troposphere	6 days to 2 weeks
Upper troposphere	2 to 4 weeks
Lower stratosphere	6 months to 1 year
Upper stratosphere	3 to 5 years
Mesosphere	5 to 10 years

This knowledge was gained at a considerable cost; it was originally assumed that fallout would descend even more slowly—that it would take seven years for half of it to reach the earth. Had that been the case, human exposure to radioactivity would have been much less because the radioactivity would have diminished by decay before the fallout could contaminate soil, water, and food. By the time we had learned the facts about fallout from the stratosphere, however, our soil and food chains were already contaminated. What we still do not know is whether man's contribution to the long-lived particulate layer in the stratosphere is—and will remain—negligible.

Particles from human activities alone may not be enough to lower the mean annual temperature of the globe, but the combination of volcanic and manmade particles may be able to do so[20]. When a volcano erupts, it sends a tremendous stream of dust and gases into the atmosphere which sometimes includes considerable water vapor. These effluents can be carried all the way to the stratosphere, where they remain long enough to affect the temperature in the stratosphere itself and in the troposphere below it. One difficulty in gauging the effect of any particular volcano is that the "normal" particulate composition of the stratosphere is not known.

Slash-and-burn agriculture is common in many parts of the world and contributes substantially to the atmosphere's particulate burden. This Applications Technology Satellite (ATS 3) photograph shows smoke from Guatemalan agricultural burning drifting north across the Gulf of Mexico (between S and T) on April 22, 1971. Another area of smoke can be seen north of Honduras (U). Smoke was visible for 25 days, beginning April 18, and may have affected the weather of the southeastern United States.

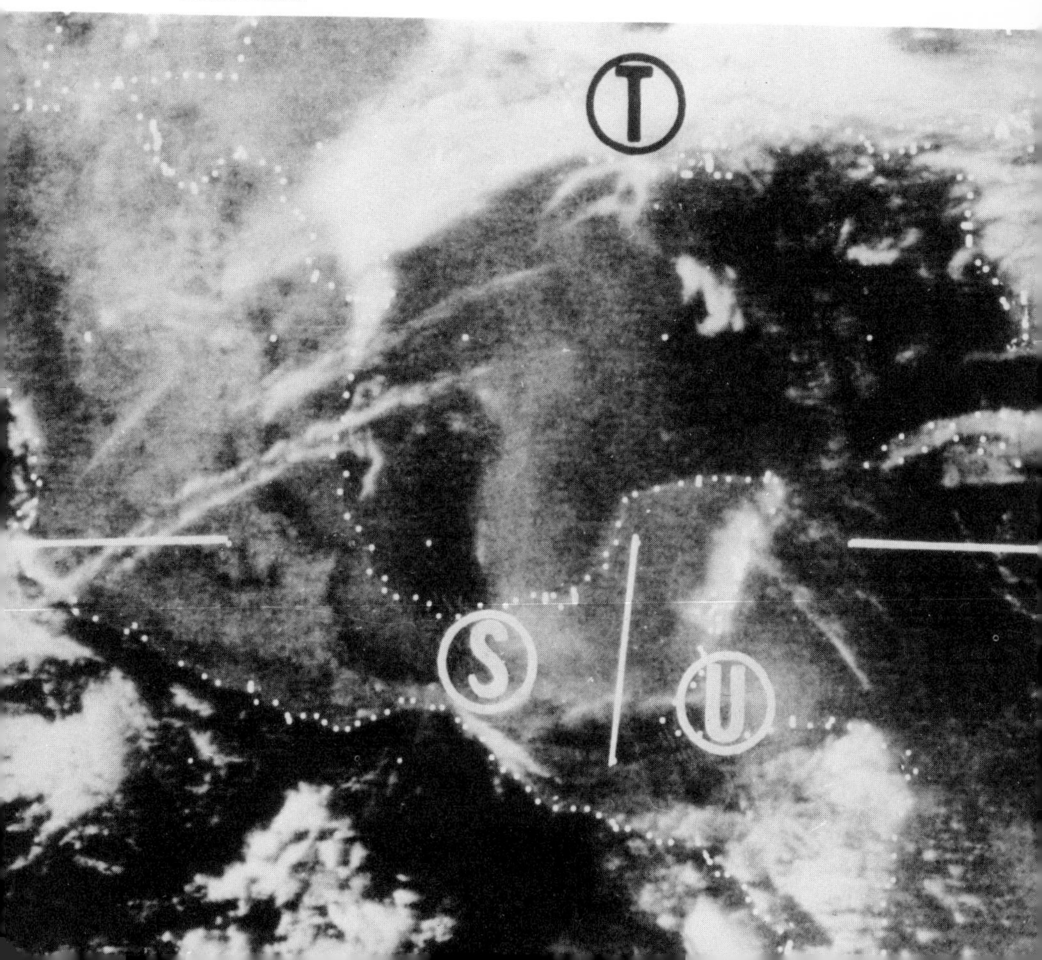

In 1963, after Mt. Agung erupted, the stratosphere was apparently *warmed* about 10.5°–12.5°F in a latitude belt about 30° wide. But in the troposphere below, the absorption of infrared radiation from the earth and the backscattering of solar radiation were apparently the dominant effects, resulting in a net *cooling* near the surface[21].

A very rough estimate puts the total annual particulate emissions of the world at about 800 million tons, equivalent to the output of a small volcano[22]. However, manmade particles are not thrust upward with volcanic force (except in the case of atmospheric nuclear explosions), and only a small fraction of the millions of tons of particulates emitted at the earth's surface reaches the upper troposphere or the stratosphere. It is the submicroscopic particles, unaffected by gravity, that stay aloft the longest and may be carried into the stratosphere, where they can remain for several years. Jet planes that travel the upper troposphere and sometimes the lower stratosphere eject particles and gases directly into those layers. Development of commercial supersonic transports has been halted in this country, but military supersonic planes are flying and the French-British and Soviet SST programs continue. It has been estimated that the peak emissions of particulates into the stratosphere from a fleet of 500 SST's would be comparable to particulates from the Mt. Agung eruption[23].

An analysis of the chemical composition of stratospheric particles has uncovered bromine in unexpectedly large quantities[24]. Bromine is used in the antiknock compounds that are added to gasoline and the possibility that this might be the source of the stratospheric bromine has stimulated a search for stable lead in the stratosphere. If lead is found, it would be a strong indicator of stratospheric pollution from the surface. A layer of sulfate and nitrate particles was also found in the stratosphere. It is generally assumed that these particles are formed from gases that are volcanic in origin, but it is not impossible that sulfur oxides and nitrogen oxides from human activities play some small part in their production.

Although aircraft appear to be the source most likely to have a significant effect on the stratosphere, there is another source of stratospheric particles and of other possible agents of climatic change which cannot be ignored. As long as there are four nuclear powers, two of them with enormous arsenals, the danger of nuclear war remains. The possibility that nuclear explosions might affect weather or climate was raised during the period of atmospheric testing. It was suggested that these effects might be created either directly by the sudden addition of such an enormous amount of energy to the atmosphere or indirectly by the emission of particles into the stratosphere or the triggering of cloudiness, precipitation, or a change in the electrical conductivity of the air.

These possibilities were all dismissed, primarily because the energy, radiation, and particles released by a single nuclear test explosion were insufficient to produce such effects[25]. In the event of a nuclear war, which might involve hundreds of times the energy, radiation, and particle production of even the largest single test, however, effects on weather and

Condensation and cloud formation from an underwater nuclear test in 1946 (like the Hiroshima bomb, about 20 kilotons). The disk-like formation on the surface of the water indicates the passage of the shock wave. In a nuclear war in which hundreds of explosions might occur, effects on the atmosphere could be considerable, but would vary with the location—on land or water—and the altitudes of the blasts.

climate cannot be ruled out. Nor can possible climatic changes be estimated, because neither the total energy nor the atmospheric layer in which the explosions would take place can be known in advance. Such effects may seem irrelevant in the face of the devastating immediate effects of nuclear war, and indeed this is true for people subjected to direct attack. But for bystanders in other countries and future generations throughout the world, climatic effects could be far from trivial.

In the troposphere particles remain a much shorter time, but dust of all kinds is being continuously pumped into this layer from jet aircraft and from fields and forests, industrial stacks, and highways below. If we are adding particles to this layer more rapidly than natural processes can remove them, we could produce a semipermanent dust screen. Comparative data over a long period of time are limited, but they are adequate to indicate that our rate of input is beginning to exceed nature's rate of removal.

An 88 percent increase in turbidity has been reported over Davos, Switzerland, a small resort town in the Alps, in about 30 years[26]. (*Turbidity* includes both the dust and the moisture in the air.) The tiny particles which serve as condensation nuclei, thus promoting the conversion of vapor to droplets, were found to increase by eight to ten times over Yellowstone Park within five years[27]. Recent evaluations of atmospheric turbidity at Mt. Olympus (Washington), Point Barrow (Alaska), and four stations in the Antarctic also suggest an increase in the earth's atmospheric particulate background[28]. A continual, even if small, interference with the amount of solar radiation reaching the earth could change the temperature on a regional and perhaps on a global scale and could also affect other aspects of the biosphere so intimately imbedded in this atmospheric layer. While plants that grow in the shade would not be affected, an increase in cloudiness or a dust-screen that reached a similar effectiveness in daytime reduction of solar radiation could affect the rate of photosynthesis in plants that require sunshine. Most crop plants grow in the sun.

Since particles can either *warm* or *cool* the atmosphere, knowing the particles are there does not tell us what effect they are having. In fact, this is a highly controversial matter, as these quotations from three papers on the subject in *Science*, published within a few months of one another, show:

> . . . the net effect of the manmade particulates seems to be that they lead to heating of the atmospheric layer in which they abound. This is usually the stratum hugging the ground. All evidence points to temperature rises in this layer, the opposite of the popular interpretations of the dust effect[29].

> . . . the present urban aerosol-surface albedo environment would produce a warming trend of solar radiation were it the only thermal process acting. On the other hand, the aerosol-albedo combination characteristic of the desert environment and prairie environment appears to produce cooling trends. . . .[30]

> . . . it is difficult to predict the rate at which global background opacity of the atmosphere will increase with increasing particulate injection by human activities. However, it is projected that man's potential to pollute will increase six- to eightfold in the next 50 years. If this increased *rate* of injection of particulate matter in the atmosphere should raise the present global background opacity by a factor of 4, our calculations suggest a decrease in global temperature by as much as 3.5 degrees K.[*] Such a large decrease in the average surface temperature of Earth, sustained over a period of a few years, is believed to be sufficient to trigger an ice age[31].

This difference of interpretation points up the need for many more actual measurements over time and indicates the complexity of the factors that have to be considered.

Whether increased particles are affecting cloud formation and precipi-

* Degree intervals on the Kelvin (K) scale are the same as they are on the centigrade scale. 0 °K is "absolute zero," −273.16 °C.

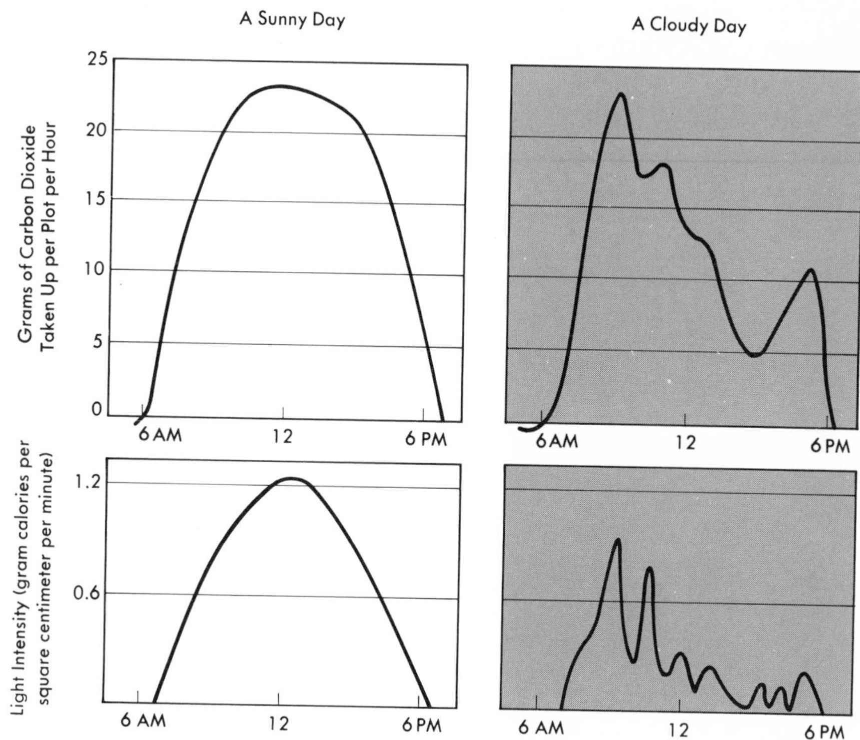

The rate of photosynthesis in many crop plants (and other plants that grow in the sun) changes dramatically with changes in light intensity. The photosynthetic rate of a plot of alfalfa in grams of carbon dioxide taken up per hour is shown as measured on a sunny day (left) and on a cloudy day (right). Pronounced increases in cloudiness or in persistence of a particulate pollution layer may therefore affect plant growth.

tation on more than a local scale is crucial to their overall climatic effect. Lead iodide has sometimes been used for seeding clouds, and it is quite possible that we have been inadvertently seeding the clouds with the lead particles from gasoline[32]. Cloud formation affects both precipitation and the atmosphere's reflectivity, and the updrafts in storm clouds can carry particles into the stratosphere.

When particles fall out, the bright, highly-reflective surfaces of snow and ice may be dimmed and, if these surfaces become dirty enough, melting can be induced. Particles from industrial pollution are carried all the way to the North Pole, as Clair Patterson found when he investigated the lead content of Arctic snow: Snow layers showed an increase of almost 300 percent in lead content between 1940 and 1965[33]. A Soviet scientist analyzed the dust content of the Caucasus glaciers and found an increase paralleling the development of industry in eastern Europe[34]. Whether this has been great enough to affect the amount of solar radiation reflected and to hasten the melting of ice and snow is not known. Decreases in temperature, whatever their cause, can set the same multiplying effects in motion that increases in temperature can—but in the opposite direction.

Finally, the physics and chemistry of particles in the atmosphere are extremely complex, with much to be learned before the whole subject of particle effect on climate can be resolved. Meanwhile, we continue to add to the annual particulate and carbon dioxide burden of the air as if all the unanswered questions were going to be answered in such a way that we would never suffer any ill effects from our own imprudence.

Whenever and however the carbon dioxide content of the earth-atmosphere system reaches an equilibrium—a point at which the amount being added to the atmosphere is equalled by the amount being withdrawn by the oceans and green plants—man has initiated a change that can be reversed only by the slow build-up of new geological deposits of carbon. The more fossil fuel we burn, the larger the amount of carbon moving through the cycle will be, and the longer it will take for the system to reach a new equilibrium.

Particulate increase, however, is at least theoretically reversible within a relatively short time. Unfortunately, particulate sources are many and various, and present control methods are ineffective in dealing with the very fine particles—precisely where the climatic danger is greatest.

The Climate of Cities

While the effect of human activities on the global climate is a controversial question, there is general agreement that these activities are now having an effect on the climate of cities. The thermal pollution of the atmosphere appears small when averaged for the whole globe or for the continents. In the areas where it is already concentrated, it looks much larger and the point when its effects could become severe seems much closer. In some places, man's activities already rival the sun in producing heat near the surface. In the summer, the amount of heat produced by combustion alone in Manhattan is one-sixth that of the solar energy reaching the ground and in winter it is two and a half times the solar energy at the surface[35].

Higher temperatures in cities are the result of surface changes and gaseous and particulate pollution, as well as combustion. Concrete, brick, and asphalt absorb and store more heat during the day than the soil and vegetation of rural and forested areas, releasing it at night; some radiant energy is converted to stored chemical energy by vegetation, also. Hard city surfaces allow rain and snow to run off rapidly, so the cooling effect of evaporation is largely lost. Temperature differences as high as 15 °F have been noted between a city and its environs on a calm, clear night when the heat island effect (page 6) is at a maximum. City size, building density, and local meteorological factors all affect the intensity of the heat island effect.

Thermal radiation emitted at the surface is re-radiated downward by the pollution layer, warming the air over the city. Unlike the self-correcting effect of natural thermal distribution, this warming of the air has a self-perpetuating tendency—it increases the stability of the atmosphere and thereby decreases the dispersion of pollutants. However, the warming effect of the pollution layer is partially offset by its cooling effect in reflecting and scattering the incoming solar radiation. In downtown Los Angeles during the months of August–November 1954, the average total daily radiation (both direct and scattered) was about ten percent below what it would have been in the absence of pollution[36]. The Smithsonian Institution recorded a 14 percent decrease in the amount of direct sunlight reaching the center of the Mall in Washington, D.C., in 1969, as compared to the early years of the century[37]. London's famous pea-soup fogs, a combination of pollution and natural fog, made visibility so low during the 1952 air pollution disaster that buses crept through the streets, led by conductors on foot, carrying flares, and people got lost only a few blocks from their homes[38]. In the subsequent air pollution control program, London has increased its winter sunshine by 70 percent and its winter visibility by 300 percent[39].

As warm air over a city rises and combines with the water vapor emitted from many industrial stacks, it tends to increase cloudiness, an effect that is further accentuated by those particles that absorb or attract moisture from the air. The additional effect of human activity became clear when it was shown that there is more rain in cities during the week than there is on weekends. From 1953 to 1967, for example, the average rainfall in Paris was 31 percent greater on weekdays than on Saturdays and Sundays[40]. However, as seems to happen so often in meteorology, the same cause can also produce the opposite effect. If there is an overabundance of moisture-attracting *(hygroscopic)* particles for the available moisture, droplets will be small; these small droplets may not coalesce into raindrops large enough to fall and therefore precipitation will be decreased.

There is some evidence that, higher up, jet contrails increase *cirrus* cloudiness—the high feathery clouds of ice crystals. Cirrus clouds reflect solar radiation and thus may have a cooling effect on the atmosphere. Precipitation may also be increased near major air routes. If further investigation confirms these effects, they would be expected to continue to increase as subsonic jet travel continues to grow.

The climatic effects of cities are not limited to the built-up areas. A plume of heat from a city may drift several miles downwind, and the smoke pall from a city has been estimated to affect an area 50 times that of the city itself[41]. Heat island effects have been noted in small towns, while industries need not be located in cities to produce increased precipitation. High rainfall in Laporte, Indiana, has been attributed to the effect of the Chicago industrial complex, 30 miles to the west[42]. Particles from industrial emissions have recently been collected in the North Atlantic, several hundred miles from the nearest city[43].

A summary of the climatic changes produced by cities follows on the next page.

Climatic Factor	Comparison with Rural Environs
Temperature	
Annual mean	1.0 to 1.5° F higher
Winter minima	2.0 to 3.0° F higher
Relative humidity	
Annual mean	6% lower
Winter	2% lower
Summer	8% lower
Dust particles	10 times more
Cloudiness	
Clouds	5 to 10% more
Fog, winter	100% more
Fog, summer	30% more
Radiation	
Total on horizontal surface	15 to 20% less
Ultraviolet, winter	30% less
Ultraviolet, summer	5% less
Wind Speed	
Annual mean	20 to 30% lower
Extreme gusts	10 to 20% lower
Calms	5 to 20% more
Precipitation	
Amounts	5 to 10% more
Days with <0.2 inch	10% more

From James T. Peterson, *The Climate of Cities: A Survey of Recent Literature*, NAPCA, Raleigh, North Carolina, October 1969, AP-59, p. 1, Table 1. (After H. E. Landsberg, "City Air—Better or Worse," *Symposium: Air Over Cities*, U.S. Public Health Service, Taft Sanitary Engineering Center, Cincinnati, Ohio, Tech. Report A 62-5.)

Giant jetports in the future may rival cities as sources of nitric oxide and particulate pollution. Because jets generally depart in an upwind direction, the jetports will be sources of constant pollution in a downwind corridor extending 20 to 40 miles[44]. Current controls cut down mainly on hydrocarbons and, to a lesser extent, on particulates; this makes jet pollution less visible, but fails to cope with the other emissions.

Changing the properties of the earth's surface changes a whole series of interactions among surfaces, solar radiation, wind, water, and living things. We are doing this most massively by city-building, but other significant factors are strip mining, removing forests, cultivating more land, building dams and creating artificial lakes, removing thousands of acres of natural cover to make way for new highway systems, and bombing and bulldozing in Vietnam on an unprecedented scale[45]. While no one small project has any appreciable effect on global climate, the scope of such projects is growing and the small ones are multiplying rapidly. In spite of their differences, all of them tend to replace natural environments with artificial ones and complex, stable systems with simple, unstable ones; most of the new surfaces these projects create exchange the moderating effects of soil and forest canopy for the inflexibility of blacktop, barren earth, or concrete.

Jet aircraft take-off is a constant source of pollution in a downwind corridor. From the giant jetports of the future, ambient levels of nitric oxide and particulates for 20 to 40 miles downwind may equal those in present urban atmospheres.

The Study of Critical Environmental Problems (SCEP) report concluded that "It is difficult to foresee any changes in global climate" as a result of changing agricultural patterns, and similarly dismissed as unimportant other surface changes with the possible exceptions of urbanization and new, very large water bodies. Yet every one of these surface changes destroys many small ecosystems, each with its characteristic microclimate. The surface's ability to absorb and reflect, the relationships between soil water and runoff and between evapotranspiration and precipitation, the air turbulence above the surface, erosion with its resultant airborne dust—all are affected. As the atmosphere is highly nonlinear in its responses, a series of small changes can have much more than an additive effect. What seem to be minor perturbations can move the system into wider and wider swings around its average until it is no longer able to maintain its former equilibrium.

Inadvertent changes in local climates raise the same kind of scientific and social questions as intentional changes do: Is added precipation in the cities depriving rural areas of needed rain? What is the effect on the water cycle of the cities' lower evaporation and more rapid runoff? At what point does the growth of cities with their increasingly warmer climates reach a point where the heat becomes unbearable? As cities grow into metropolitan areas and metropolitan areas coalesce into megalopolises, how will their regional climatic changes affect large-scale atmospheric motion and heat balance? What, if anything, should be done to control surface changes? To control aircraft emissions? To change our cities—to limit their growth and slow their accelerating energy use? While the answers to some of the scientific questions are still tentative and confusing, we do appear to be changing the atmosphere's composition and unbalancing the earth's energy budget, thus tampering with the atmosphere's complex radiation fluxes and energy-matter relationships.

Changing Chemical Relationships in the Atmosphere

The photochemical smog that has become such a well-known and thoroughly disliked aspect of city life has introduced us all to the notion that people's activities are affecting the chemistry of the atmosphere. But these chemical changes are not limited to urban smog or to gaseous emissions; particles, too, undergo complex changes in the atmosphere through coagulation with other particles, gas reactions on their surfaces, and cycles of condensation and evaporation in water droplets.

Until quite recently, there was concern that carbon monoxide might be accumulating in the air. Man is adding millions of tons to the atmosphere annually and a buildup would be very serious because of the toxicity of this gas. However, experiments have now demonstrated that there is a natural removal mechanism in the soil, probably the activity of microorganisms, which counteracts this input. While the carbon monoxide content of the atmosphere can rise near traffic and other combustion sources and can remain high, especially in city centers where there is little exposed soil surface, these experiments show that, if the soil samples tested prove to be typical of average soil activity in this regard, all the carbon monoxide we are presently adding to the air can be rapidly removed[46].

The fear that we might deplete our supply of oxygen has also been laid to rest. Although combustion not only produces carbon dioxide, but uses up oxygen, estimates indicate that burning all the known reserves of fossil fuel would leave us with still-breathable air. The oxygen in the atmosphere might be reduced by two percent[47] or even as little as 1.2 percent[48].

We are interfering with the production of oxygen on land by city-building and other acitivities which limit the land devoted to green plants. We are interfering with the production of oxygen at sea (where marine phytoplankton are responsible for about 60 percent of global photosynthesis) by injecting chemical pollutants into marine environments. DDT can reduce photosynthesis in some coastal and oceanic algae[49]. These interferences with the photosynthetic process may create a genuine ecological problem because green plants provide food for all other forms of life, but this problem is not the reduction of the oxygen content of the air. The oxygen produced by green plants is consumed almost entirely each year by animals and bacteria, so that oxygen is being added to and extracted from the atmosphere at very close to the same rate. If more oxygen is added, more is taken away and if less is added, less is taken away, so that the oxygen content of the atmosphere remains at 20.946 percent.

Calculations of the possible depletion of the ozone shield that protects the earth from ultraviolet radiation are less reassuring. Photochemical reactions of the ozone with water vapor or with the nitrogen oxides emitted by supersonic transports are the source of this concern. Although most calculations indicate that the amount of water vapor from a fleet of 500 SST's would reduce the ozone by only about four percent, which is no more than present natural variations, there are a number of uncertainties involved. The effect of nitrogen oxides on the ozone layer may present a more serious problem.

In nature, there are relatively small amounts of the molecular forms of nitrogen: nitrate, nitrite, and several gaseous nitrogen oxides in which nitrogen and oxygen are combined. As nitrogen passes through the successive stages of its cycle in the biosphere, it rarely finds itself combined solely with oxygen for any long period, except in inorganic mineral deposits. Because the nitrogen gas that makes up nearly 80 percent of the air is very stable chemically, it will not react with oxygen to form nitrogen oxides (and eventually nitrate) except at temperatures so high that under natural conditions they are produced only by lightning. At the temperatures reached in the boilers of fossil-fueled power plants, internal combustion and jet engines, industrial processes, and large incinerators, however, the normally stable nitrogen reacts with oxygen to form nitric oxide which is converted in the air to nitrogen dioxide. The nitrogen oxides normally present in the stratosphere may play a role in limiting ozone, and the introduction of the amount of nitrogen oxides presently expected from SST exhaust could initiate a series of chemical reactions which would reduce the ozone in the atmosphere by as much as half, permitting ultraviolet radiation to penetrate the ozone shield and reach the earth[50]. We need to know more about the actual emissions of the SST and their effect on stratospheric chemistry before these disquieting questions can be resolved.

The behavior of nitrogen oxides in the lower atmosphere is better understood. They play an important role on sunny days in the formation of photochemical smog.

Photochemical Smog

The sun generates the power for the "chemical factory" of the urban atmosphere. The raw materials are nitrogen oxides, oxygen and water vapor in the air, the hydrocarbons, sulfur oxides, and all of the other contaminants that are added in each city. The output is photochemical smog.

Smog has been studied most in Los Angeles, where it first became a severe problem. While automobile emissions are much the same everywhere, there is a much greater input of sulfur oxides in cities like Chicago and New York and of more exotic chemicals in chemical manufacturing centers like St. Louis and the Kanawha Valley of West Virginia. However, while the mix of chemicals in the photochemical system is not the same everywhere, the broad outlines of the process and some of the major secondary contaminants probably vary little from city to city. A great deal of experimentation over the last 20 years has established the basic phenomena involved and has reconciled most observations with well-known chemical principles. It has long been evident, for example, that hydrogen peroxide should be among the oxidants in smog and this compound has recently been identified in Los Angeles smog. It is now possible to simulate the development of photochemical smog in the Los Angeles atmosphere by computer, using traffic survey information to estimate the rate of pollution in various parts of the city[51].

Nitrogen dioxide initiates the process that produces photochemical smog. Brownish in color, it is able to absorb light over the whole visible and ultraviolet range of the solar spectrum in the lower atmosphere, thus capturing the sun's energy and thereby initiating the chain of chemical events that produces new chemical compounds. Activated by the absorbed light, nitrogen dioxide is converted to nitric oxide and some highly reactive atoms of oxygen. In turn, the oxygen atoms react with ordinary oxygen in the air to form ozone. But this chain of chemical events leads nowhere: There is no buildup of ozone or other oxidants unless hydrocarbons or similar organic compounds are present to complete the process. And some of these compounds are much less effective than others in this respect; in particular, methane, the most abundant hydrocarbon in the atmosphere, has no effect at all. However, some of the hydrocarbons and other compounds found in automobile exhausts are very effective in promoting the chain.

The hydrocarbons constitute a very extensive group of chemical compounds made up of atoms of carbon and hydrogen arranged into various structures, some of them quite complex. For convenience, they are classified in various types. The major varieties found in gasoline are known as paraffins, olefins, aromatics, and naphthenes. These types differ in both the ease in which they can be manufactured into secondary contaminants and the extent to which they accelerate the conversion of nitric oxide to nitrogen dioxide and the subsequent buildup of ozone and other oxidants. The most reactive known hydrocarbons are among the olefins, but the others are utilized to some extent in the photochemical process. Although automobile exhausts contribute a large share of the hydrocarbons in the

The daily course of photochemical smog: Before dawn, when the city is quiet and traffic is at a minimum, concentrations of the primary contaminants—carbon monoxide, nitric oxide, and hydrocarbons—increase slowly in the absence of wind. The rising of the sun and the burgeoning of morning traffic speed their accumulation. About two hours after dawn, nitrogen dioxide begins to be generated at a substantial rate. Then, usually within one to two hours, the nitric oxide is reduced to low values because it has been converted to nitrogen dioxide; at this time the nitrogen dioxide concentration has reached its peak. The disappearance of nitric oxide coincides with the appearance of ozone. (The two cannot coexist in appreciable concentrations because they react rapidly with one another, yielding nitrogen dioxide.) Ozone reaches its maximum sometime after noon and then gradually begins to decline. The concentration of nitrogen dioxide declines from its peak as the ozone builds up, and is usually negligible by late afternoon. The afternoon traffic peak injects an additional burden of nitric oxide into the atmosphere, which scavenges the remaining traces of ozone by early evening. The primary pollutants then reaccumulate at a decreasing rate for the remainder of the night.

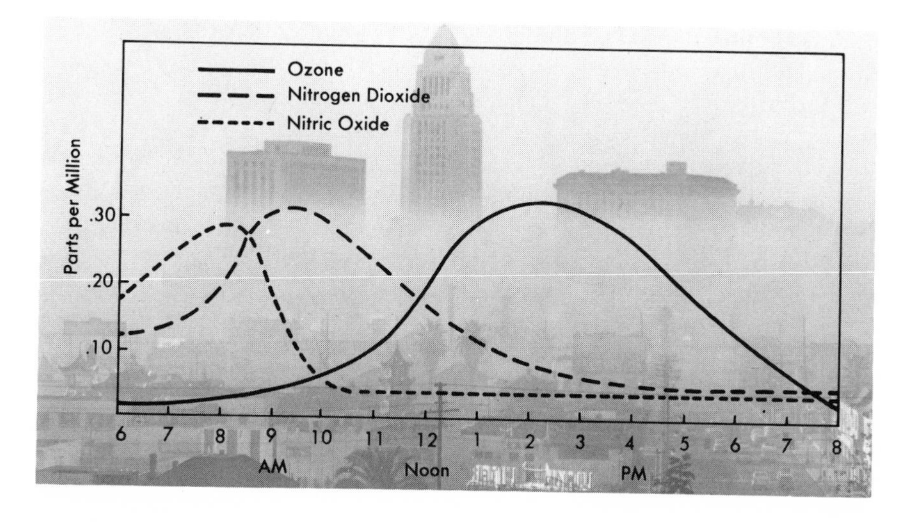

city atmosphere, some are also emitted from industrial and incinerator stacks.

The volatile hydrocarbons not only promote the production of nitrogen dioxide and ozone but also react with these gases to form what is now familiarly, though not affectionately, called PAN (peroxyacetyl nitrate)—an eye-stinging, vegetation-destroying compound. Other products of the urban chemical factory are the aldehydes—formaldehyde, acrolein, and others—which represent one state in the oxidation of hydrocarbons. The ultimate fate of these hydrocarbons appears to be oxidation to carbon monoxide or carbon dioxide and water. However, there are also particles of soot in the air which are agglomerations of hydrocarbon molecules. The daily pattern of events on a typical smog day is shown in the above graph. The story is repeated with some variations on every calm, sunny day in every industrial city with heavy automobile traffic, and especially when an inversion keeps a "roof" on the chemical factory.

The first efforts to reduce automobile emissions increased fuel combustion through greater air intake, thereby decreasing hydrocarbons and carbon monoxide, but *increasing* nitrogen oxides. The rapid reactions between ozone and nitrogen oxides show why the toxic ozone, although harmful to both plants and people while it lasts, does not build up over time to lethal concentrations. This relationship also suggests why reactions between ozone and nitrogen oxides in the stratosphere may deplete the ozone shield.

While the smog potential depends primarily upon the levels of sunlight, oxides of nitrogen, and reactive hydrocarbons, laboratory studies reveal that other contaminants such as sulfur dioxide, carbon monoxide, and particulate matter have relatively minor, but still measurable, effects. Carbon monoxide, for example, seems to accelerate the formation of nitrogen dioxide and ozone[52]. Changes of temperature and humidity also affect the rate of oxidant production to some degree.

The possibility that a reduction in sulfur oxides in the New York atmosphere had *increased* photochemical smog was suggested by Austin Heller, then Commissioner of Air Resources, when he noted that the former was dropping while the latter was rising. Theoretically this is possible, since sulfur dioxide is a reducing agent. (It is added to some foods to counter the oxidizing which results in spoilage.) Therefore it is conceivable that the sulfur dioxide in the air could counter the oxidizing of nitric oxide and hydrocarbons which is such an important part of the photochemical system. It is interesting to note that in Chicago, where sulfur pollution was high, the oxidant level dropped slightly during Episode 104. However, high oxidants appeared as often in association with high sulfur dioxide as with low sulfur dioxide in a Philadelphia study and no definite trend toward increasing oxidants with decreasing sulfur dioxide was found[53]. A laboratory investigation subsequently showed that in a dry system the ozone concentration increases with increasing sulfur dioxide, but at a relative humidity of 65 percent the ozone concentration decreases with increasing sulfur dioxide[54].

Study of the complex chemical reactions in the photochemical system as a whole have been frustrated to some extent by the impossibility (at least thus far) of identifying some of the components of the polluted air at low concentrations and also by the fact that isolating others for study artificially simplifies the system.

[1] U.S. Committee for Global Atmospheric Research Program, Division of Physical Sciences, National Research Council, *Plan for U.S. Participation in the Global Atmospheric Research Program,* National Academy of Sciences, Washington, D.C., 1969.

[2] Ibid., p. 5.

[3] William D. Sellers, "A Global Climatic Model Based on the Energy Balance of the Earth-Atmosphere System," *Journal of Applied Meteorology,* June 1969, 8:392–400 and "Comments" by M. I. Budyko in the same journal, April 1970, 9:310.

[4] Frederick Sargent, II, "Taming the Weather, A Dangerous Game," *Environment,* May 1967, Vol. 9, No. 5, 81–88.

[5] "A Tropical Garden or a New Ice Age," *Nuclear Information* (now *Environment*), Vol. 1, No. 4, February 1959, 3. The theory was proposed by Maurice Ewing and William L. Donn in "A Theory of Ice Ages," *Science,* June 15, 1956, 123:1061–66.

[6] Allen L. Hammond, "Weather Modification: A Technology Coming of Age," *Science,* May 7, 1971, 172:248–49.

[7] H. Peter Metzger, "The Day It Rained in Rapid City," *New York Times* News Service story in *St. Louis Post-Dispatch,* September 19, 1972.

[8] Gordon J. F. McDonald, "How to Wreck the Environment," in *Unless Peace Comes,* Nigel Calder (ed.) (New York: The Viking Press, 1968), pp. 165–83.

[9] Deborah Shapley, "Rainmaking: Rumored Use over Laos Alarms Arms Experts, Scientists," *Science,* June 16, 1972, 176:1216–20, and "Rainmaking: Stockholm Stand Watered Down for Military," *Science,* June 30, 1972, 176: 1404. See also Seymour M. Hersh in the *New York Times,* July 3, 1972, 1–2.

[10] Hermann Flohn, *Climate and Weather* (New York: McGraw-Hill Book Company, 1969), pp. 236–88.

[11] In West Pakistan, for example, 100,000 acres of rich farmland were lost every year during the 1960s from waterlogging and salinity after the initiation of a massive irrigation project. (Martin Adeney, "Food Yields from Salty Lands," *New Scientist,* reprinted in *Hunger,* Scientists' Institute for Public Information Workbook, New York, 1970.) The Aswan Dam has interfered with the Mediterranean sardine fisheries, increased the incidence of schistosomiasis (a parasitic disease caused by blood flukes), and blocked the deposit of nutrient-rich silt in the Nile Valley. (Harmon Henkin, "Side Effects," *Environment,* January–February, 1969, 11:32–33).

[12] In an article in *Science,* April 24, 1970, 168:473–75, R. Cecil Gentry of the National Hurricane Research Laboratory, Environmental Science Services Administration, describing experiments in seeding Hurricane Debbie in August 1969, stated: "That Hurricane Debbie decreased in intensity following multiple seedings on 18 and 20 August is well established. What we do not know is whether the decrease was caused by the seeding or by natural changes in the hurricanes." He added that ". . . the 1969 experiments suggest strongly that modification was accomplished."

[13] Sargent, "Taming the Weather" (see [4]).

[14] A new book on weather modifications is available: *Inadvertent Climate Modification,* Report of the Study of Man's Impact on Climate (SMIC) (Cambridge: MIT Press, 1971).

[15] Warren Washington, "On the Possible Uses of Global Atmospheric Models for the Study of Air and Thermal Pollution," in *Man's Impact on the Climate,* W. H. Mathews, W. W. Kellogg, and G. D. Robinson (eds.) (Cambridge: MIT Press, 1971).

[16] Sellers, "A Global Climatic Model" (see [3]). Sellers uses the figure for 1966 of 0.02 kilolangleys per year for the quantity of energy used by man and converted into heat. (A *langley* is a unit of heat related to a unit of surface. One langley equals one gram-calorie per square centimeter. A *kilolangley* is a thousand langleys. The amount of solar radiation available for absorption within the atmospheric system is about 0.5 langleys per minute, which is 720 langleys per day or 2628 kilolangleys per year.) Sellers projects an increase from man's activities to 50 kilolangleys per year. In order to roughly project how long it would take to reach this quantity a table from a United Nations report on world energy was used. This table gives an assumed yearly world increase in thermal waste energy of 5.7 percent. If this rate of increase were to continue, the world would reach Seller's figure in about 330 years. Washington (reference[5]) used the addition of 50 langleys per day (18.25 kilolangleys per year) of heat from conversion of energy by man, an amount the world may reach in about 115 years. A brief discussion of computer simulation of the atmosphere as a predictive tool is included in *Man's Impact on the Global Environment,* the report of the Study of Critical Environmental Problems (SCEP) (Cambridge: MIT Press, 1970), pp. 78–82. The UN energy table will be found on p. 64 of this report.

[17] Some other sources of increased carbon dioxide were indicated by a study of carbon 14, whose long half-life (about 5000 years) can help to date carbon dioxide in the air and therefore suggest its course. The carbon in coal, for example, is so old that the carbon 14 has almost completely decayed, while this is not true of the carbon in trees. Roger Revelle and H. E. Suess in "Carbon Dioxide Exchange Between Atmosphere and Ocean and the Question of an Increase of Atmospheric CO_2 During the Past Decade" (*Tellus,* February 1957, 9:18–27) found that some of the carbon dioxide in the atmosphere must come from the burning of materials more recent in origin than coal deposits—forest fires, slash-and-burn agriculture, or the slow oxidation of peat bogs and soil humus. E. S. Deevey, Jr. in "Bogs" (*Scientific American,* October 1958, 115–21) believes the last of these is an important source. However, there is general agreement that combustion is an important factor in the increased amount of carbon dioxide in the atmosphere.

[18] *Man's Impact on the Global Environment,* SCEP report, p. 12 (see [16]).

[19] S. I. Rasool and S. H. Schneider, "Atmospheric Carbon Dioxide and Aerosols: Effects of Large Increases on Global Climate," *Science,* July 9, 1971, 173: 138–41.

[20] Reid Bryson, "All Other Factors Being Constant . . .," *Weatherwise,* April 1968, 56–62.

[21] SCEP Report, p. 104 ([16]).

[22] Eugene K. Peterson, "The Atmosphere: A Clouded Horizon," *Environment,* April 1970, Vol. 12, No. 3, 35.

[23] SCEP Report, p. 107 ([16]).

[24] SCEP Report, p. 58 ([15]). The study was reported by R. D. Cadle et al. in "The Chemical Composition of Aerosol Particles in the Tropical Stratosphere," *Proceedings of the American Meteorological Society Symposium on Tropical Meteorology* (unpublished).

[25] Samuel Glasstone (ed.), *The Effects of Nuclear Weapons,* U.S. Department of Defense and the Atomic Energy Commission, USGPO, April 1962. Revised edition, February 1964, pp. 84–86.

[26] R. A. McCormick and J. H. Ludwig, "Climate Modification by Atmospheric Aerosols," *Science,* June 9, 1967, 156:1358–59.

[27] Vincent J. Schaefer, "The Inadvertent Modification of the Atmosphere by Air Pollution," *American Meteorological Society Bulletin,* April 1969, 50:199.

[28] William H. Fischer, "Atmospheric Aerosol Background Level," *Science,* February 26, 1971, 171:828–29.

[29] Helmut E. Landsberg, "Man-Made Climatic Changes," *Science,* December 18, 1970, 170:1267–74.

[30] Marshall Atwater, "Planetary Albedo Changes Due to Aerosols," *Science,* October 2, 1970, 170:64–66.

[31] Rasool and Schneider, "Atmospheric Carbon Dioxide and Aerosols," p. 141 (see [15]).

[32] Schaefer, "The Inadvertent Modification of the Atmosphere" (see [27]).

[33] Clair C. Patterson, with Joseph P. Salvia, "Lead in the Modern Environment," *Scientist and Citizen* (now *Environment*), April 1968, 10:66–79.

[34] F. F. Davitaya, quoted by James T. Peterson in *The Climate of Cities: A Survey of Recent Literature,* NAPCA, Raleigh, North Carolina, October 1969, AP-59.

[35] R. D. Bornstein, "Observations of the Urban Heat Island Effect in New York

City," *Journal of Applied Meteorology,* August 1968, 7:575. Quoted by Peterson in *The Climate of Cities* (see [34]).

[36] N. A. Renzetti, *Air Pollution Foundation Report,* 1955, 9:200. Quoted by A. J. Haagen-Smit and Lowell G. Wayne in "Atmospheric Reactions and Scavenging," Arthur C. Stern (ed.), *Air Pollution,* Vol. I, second edition (New York: Academic Press, 1968), pp. 154–55.

[37] *Smithsonian,* May 1970, 1:2.

[38] William Wise, *Killer Smog* (New York: Rand McNally & Co., 1968).

[39] James T. Peterson, *The Climate of Cities,* p. 33 (see [34]).

[40] Landsberg, "Man-Made Climatic Changes," p. 1272 (see [25]).

[41] Ibid.

[42] Stanley A. Changnon, Jr., "Recent Studies of Urban Effects on Precipitation in the United States," *American Meteorological Society Bulletin,* June 1969, 50:411.

[43] D. W. Parkin, D. R. Phillips, and R. A. L. Sullivan, "Airborne Dust Collections Over the North Atlantic," *Journal of Geophysical Research,* March 20, 1970, 7:1782–92.

[44] James A. Fay, "Air Pollution from Future Giant Jetports," paper presented at the 63rd Annual Meeting of the Air Pollution Control Association, St. Louis, Missouri, June 1970.

[45] A bomb called the "Daisy Cutter" can destroy a 300-acre area (although it can kill or injure life in a 1746-acre area). These bombs have been used extensively in Vietnam since early 1970. The bulldozers clear more than 1000 acres a day. Defoliation, bombing, and bulldozing had destroyed at least 12 percent of the forest cover of South Vietnam by late 1971. (E. W. Pfeiffer and Arthur W. Westing, "Land War," *Environment,* Vol. 13, No. 9, 2–15.

[46] Robert E. Inman, Royal B. Ingersoll, and Elaine A. Levy, "Soil: A Natural Sink for Carbon Monoxide," *Science,* June 18, 1971, 172:1229–31.

[47] Wallace S. Broecker, "Man's Oxygen Reserves," *Science,* June 26, 1970, 168:1537–38.

[48] Eugene K. Peterson, "The Atmosphere: A Clouded Horizon," *Environment,* April 1970, Vol. 12, No. 3, 32–39, and subsequent correspondence, December 1970, Vol. 12, No. 10, 44–45.

[49] Charles F. Wurster, Jr., "DDT Reduces Photosynthesis by Marine Phytoplankton," *Science,* March 29, 1968, 159:1474–75.

[50] Harold Johnston, "Reduction of Stratospheric Ozone by Nitrogen Oxide Catalysts from Supersonic Transport Exhaust," *Science,* August 6, 1971, 173:517–22.

[51] Lowell G. Wayne et al., "Modeling Photochemical Smog on a Computer for Decision-Making," paper presented at the 63rd Annual Meeting of the Air Pollution Control Association, St. Louis, Missouri, June 1970.

[52] Karl Westberg, Norman Cohen, and K. W. Wilson, "Carbon Monoxide: Its Role in Photochemical Smog Formation," *Science,* March 12, 1971, 171: 1013–15.

[53] Aubrey P. Altshuller, "Relationships Among Pollutants Undergoing Atmospheric Reactions in CAMP Cities in the U.S.," paper presented at the 63rd Annual Meeting of the Air Pollution Control Association, St. Louis, Missouri, June 1970.

[54] Aubrey P. Altshuller and Joseph J. Bufalina, "Photochemical Aspects of Air Pollution: A Review," *Environmental Science and Technology,* January 1971, 5:39–64. The report on humidity mentioned by Altshuller and Bufalini is "The Effect of Water Vapor on the Oxidation of 1-butene and NO in the Photochemical Smog Reaction" by W. E. Wilson and A. Levy, which was presented at The American Chemical Society Meeting, Minneapolis, Minnesota, April 1969.

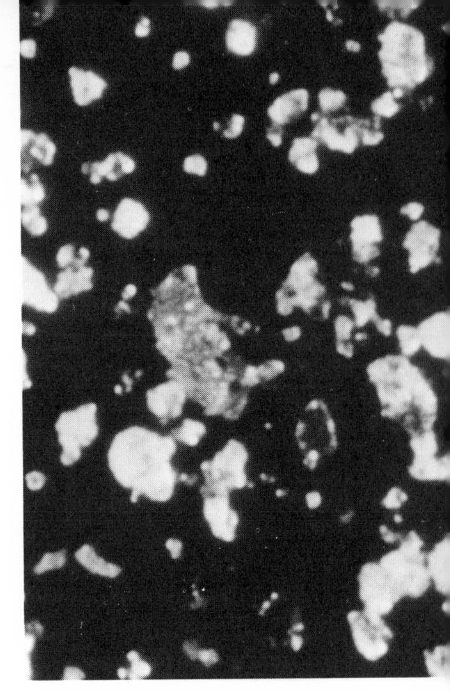

Fallout

"What goes up, must come down," as every child knows. Yet we have been slow to learn the fate of the many kinds of waste fouling the air and eventually reaching other parts of our environment. Many elements with a minor role, or no role at all, in natural life processes play a major role in our urban civilization. Produced over millions of years and held in the rocks of the earth's crust, these elements are released slowly by weathering under natural conditions, but we are now mobilizing them rapidly, which presents problems both of resource depletion and air pollution. Some do not volatilize at ambient air temperatures and are therefore present in the air only rarely and in immeasurably small amounts under natural conditions. Reduced to small particles or vaporized at the high temperatures characteristic of our energetic technology, they are emitted into the air, but eventually fall out over water and land. We are also producing synthetic materials and radioactive waste. Neither would appear in the environment without man's intervention, and the ability of natural processes to accommodate them is open to serious question.

Ultimately, people directly inhale a portion of the air's gaseous and particulate contaminants and, in addition, ingest some of the contaminants that fall out on soil and water and are taken up by other forms of life. The direct effects of air pollution on human health will be dealt with in the next chapter. Here we will concern ourselves with the range of contaminants and their effects on various aspects of the environment that support human life.

Vegetation acts as a sink for some gaseous pollutants[1]. Plants remove

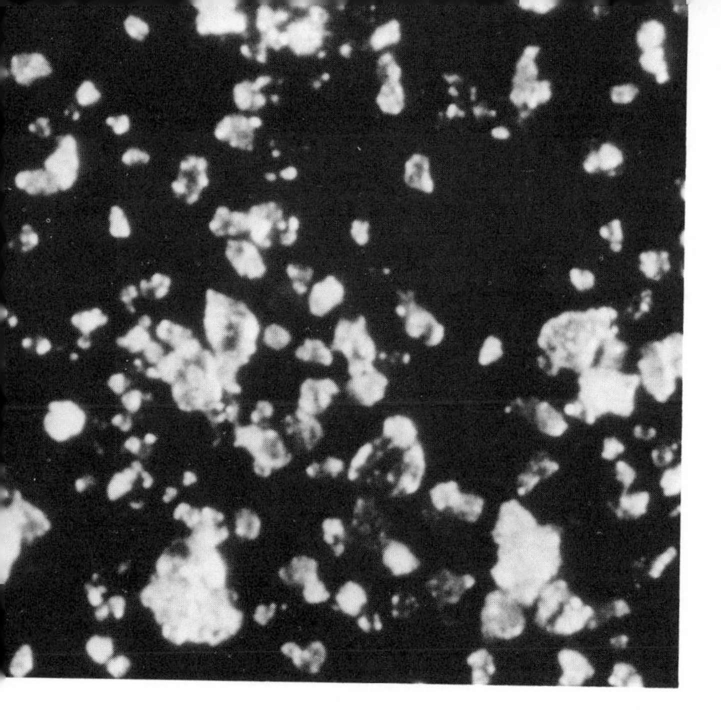

hydrogen fluoride, sulfur dioxide, and some components of photochemical smog from the air, thus helping to purify it—especially during the growing season. Since one of the functions of a leaf is the removal of carbon dioxide from the air, the same highly specialized structure can also remove other gases. However, these other gases are not part of the plant's natural chemistry and can damage or even kill it. High concentrations of ozone and chlorine cause the stomatal openings on alfalfa leaf surfaces to close partially, so that the gases can no longer be taken in at the same rate. Some trees are so severely damaged by pollution that they lose their leaves and are no longer able to function as air purifiers. After prolonged exposure, the most sensitive species may be killed.

Photochemical smog retards growth, reduces crop yield, and renders plants more susceptible to insect attack[2]. The sensitive ponderosa pines of the San Bernardino Mountains in southern California, after being injured by smog, were more readily attacked by bark beetles than were healthy trees[3]. Stands of ponderosa pine in the Sequoia National Forest in the central part of the state and in the Santa Cruz Mountains in the northern part have also been injured by smog[4]. Experiments at the University of California at Riverside found that lemon trees in the open, smoggy air produced 43 percent less fruit than trees protected by plastic greenhouses and supplied with filtered air. Early loss of leaves, smaller fruit, and generally poorer growth have been noted throughout the citrus groves of the Los Angeles area. Crop losses from air pollution damage nationally run into millions of dollars a year, some of it from the PAN and ozone in photo-

Thousands of the beautiful ponderosa pines of California, from the San Bernadino Mountains of the South to the Santa Cruz Mountains of the North, have been injured or killed by smog. Some trees, weakened by smog, have become vulnerable to insect attack. Of the two trees in this picture, one is dying from smog damage, while the other appears to be still healthy.

chemical smog, some from sulfur fumes and other pollutants. All the air pollutants that are toxic to plants have not been identified. Plant injury has been found which appears to be attributable to photochemical air pollution, but which is different from damage caused by PAN, ozone, or other known photochemical toxicants[5].

Chemical Fallout

Sulfur. Stored sulfur is being released into the air along with the stored carbon from fossil fuels. A recent assessment of the sulfur cycle concluded that "Man is now contributing about one half as much as nature to the total atmospheric burden of sulfur compounds, but by 2000 AD he will be contributing about as much and in the Northern Hemisphere alone he will be more than matching nature."[6]

Since deposits of sulfur are plentiful, there has been no concern about the wastefulness of this process, and sulfur mining continues as millions of tons of waste sulfur are dumped into the air. In 1966, the United States produced more than eight million long tons of sulfur, of which about 18 percent was a by-product of natural gas production[7]. At the same time, sulfur emissions from industrial and power plant stacks were estimated at 23 million long tons. There are processes available for the recovery of sulfur from coal and oil before burning and from stack gas after burning, but it is *cheaper* to mine elemental sulfur. Interestingly, however, the reserves of native sulfur amount to only about 210 to 230 million long tons. Another 250 million is in sulfide ores, petroleum, oil shale, and natural gas, but by far the most—an estimated 21,000 million long tons—is in coal deposits. As we mine and burn the coal, this sulfur is going up the stacks[8]. In the natural cycle, some sulfur eventually reaches the ocean bottom where it becomes irretrievable. Man is accelerating this process.

The sulfur emitted as sulfur dioxide is further oxidized in the air to sulfur trioxide, which reacts with water vapor to form sulfuric acid mist. The sulfuric acid reacts with other components of the urban atmosphere to form sulfate particles. Sulfur pollution also appears as hydrogen sulfide from paper mills and some other industrial sources, and as mercaptans, organic sulfur compounds which are also in paper mill effluent. The end product of sulfur in the air is the sulfate particles which fall out or are washed out in rain and snow. However, sulfur pollution may come in contact with plants at earlier stages in this series of changes.

When the sulfur oxides come in contact with living tissue, they can cause damage. Plants have been fumigated in the laboratory with sulfur dioxide, alone and in combination with other pollutants, and different pathologies have been observed. Careful examination of plants and trees in the field then reveal whether they have been damaged by sulfur oxides, and what other pollutants may be implicated. Levels of sulfur dioxide formerly thought to be too low to cause acute plant injury, when combined with ozone (also at subacute levels) have been found to injure tobacco plants[9].

Severe damage has been observed in the vicinity of smelters, where heavy sulfur emissions are common. Ten years of observation in the forest area near Sudbury, Ontario, smelters showed severe tree damage in a zone of about 720 square miles. Annual mortality of white pines was higher there than in forests farther from the smelters and growth of living pines was lower. The loss in white pine alone was estimated at $117,000 annually[10].

Power-plant effluents are also responsible for sulfur damage to trees and sometimes for sulfur-fluoride damage, as some coal is high in fluoride. A series of laboratory experiments and field observations with sulfur dioxide and fluoride have been carried out by Clarence C. Gordon. His investigation of Christmas tree plantations in the vicinity of the Virginia Electric Power Company found severe sulfur damage close to the plant, gradually lessening as the distance from the plant increased. He predicted that the area downwind (two miles north) of the plant would become essentially barren within two years[11]. More surprising than the damage

These eastern white pines were not killed by smog, but by another type of air pollution—the sulfurous emissions of a smelter at Sudbury, Ontario.

near the giant utility, however, was Gordon's observation that lesser symptoms of sulfur damage were appearing throughout the forests of the industrial Northeast. He has traced the source to rain that has been rendered acid by sulfur emissions, a phenomenon that has also been observed in the Scandinavian countries where the acidity of precipitation is increasing by an estimated one to two percent annually[12].

Strongly acid rain, which contains flyash as well, has been observed in Scandinavia following stagnant air situations over Central Europe. Apparently the pollutants accumulate in that area and are then transported over long distances by the general circulation of the atmosphere. The U.N. Office for Economic Cooperation and Development is planning a study which will seek to link sources, meteorology, and acidity in precipitation. It is already clear that the acid rain has increased the acidity of rivers and lakes in Norway, with effects varying from considerable damage to complete destruction of the fish life in most rivers and thousands of lakes in the southern part of Norway. When the pH drops below 5, the salmon disappear; when it falls below 4.5, the trout disappear also.*

---

*pH is a measure of acidity, with 7 being neutral, anything below that, acid, and anything above it, alkaline. The scale is logarithmic, so a pH of 4 is ten times more acid than a pH of 5.

It has long been known that heavy sulfur fumes from industrial plants could render the soil in the vicinity more acid, but it is only recently that precipitation in areas at a considerable distance from utilities or other industrial sources has been found with a pH as low as 3.5 to 3[13]. Plants differ in their adaptability to soil acidity, but most plants grow best under slightly acid conditions in soil with a pH of 6.0 to 6.9. A study of a watershed in New Hampshire found that more sulfate was coming into the watershed

Changes in the acidity of precipitation over northern Europe from 1956 to 1966.

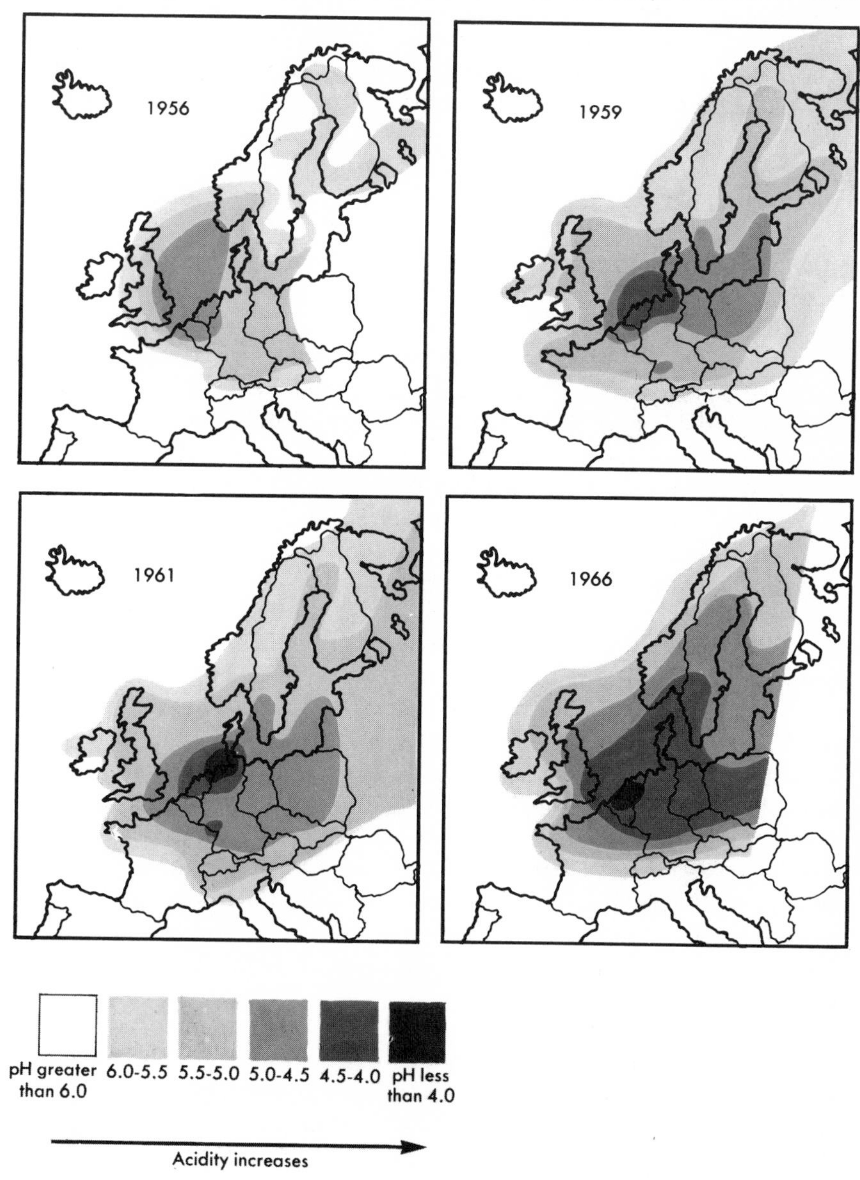

than was being removed in the streams[14]. If this is a widespread phenomenon, it indicates the possibility of a slow, but significant, buildup in soil acidity. Soil that is too alkaline might benefit from added sulfate, while soil that is already acid might suffer in terms of its ability to nourish plant growth. Changes in soil pH, like changes in temperature and moisture, can have far-reaching effects on soil microorganisms and on all the complex interactions of soil nutrition and plant life.

Sulfur oxides are also among the pollutants responsible for extensive damage to materials, at considerable cultural and economic cost. Along with other acid gases and sticky particulates, they have been implicated in the highly publicized damage to Greek and Italian art treasures. After many centuries of slow aging, paintings, statues, and whole buildings like the Acropolis are suffering accelerated destruction in twentieth-century air.

The celebrated self-cleansing properties of the air may indeed be removing the sulfur from the air, but at a cost that is just beginning to appear.

Nitrogen. The nitric oxide and nitrogen dioxide in the lower atmosphere are further oxidized to nitrogen pentoxide; combined with atmospheric moisture, this turns to nitrate and is brought down in rain and snow, where it can fertilize crops and intensify the overfertilization of our already burdened waters. The nitrogen for most of the chemical fertilizer now being used is fixed industrially from nitrogen in the air. Although the amount has been doubling about every six years, it is still negligible in comparison to the enormous amounts of nitrogen in the air[15]. However, in the other parts of the cycle, as nitrogen passes through soil, plants, and water, the imbalances introduced by this industrial fixation of nitrogen are becoming a serious problem[16].

Fifty Other Elements. Two scientists, K. K. Bertine and Edward D. Goldberg, recently estimated the amounts of sulfur and 50 other elements released to the atmosphere from combustion. Many of these have been measured in air; others have not, but their presence can be assumed because they exist as impurities in coal and oil that are released in burning[17]. In addition, when these elements are mined, transported, processed, used, and discarded, some may be lost to the air at each step of the way; other elements are present in soil and appear in airborne dust—some naturally and unavoidably, some as a result of the removal of protective vegetation. Some of these elements are essential to human health in trace amounts, but in sufficiently high doses all heavy metals and some other elements are toxic.

Not a great deal is known about the effects of these elements on various aspects of the environment as they fall through the air, are incorporated in soil, suspended or dissolved in water, and finally carried by rivers to the oceans. In the natural sedimentary cycle they are eroded from rocks by the action of wind and water, leached from soil, carried to the sea, and renewed in the earth's crust by volcanic action. The atmosphere plays a relatively small part in these natural cycles. Although great quantities of gases and particles shoot out of a volcano and return to the earth through the atmosphere, most of the natural cycle takes place at low temperatures and only a few of the elements are volatilized. But in the movement of these elements initiated by man, the atmosphere plays a larger part.

We can neither add to nor subtract from the total amount of material in the earth-atmosphere system; in that sense, a material can never be lost. We can move it around, change its form, concentrate it, or disperse it. To the extent that we disperse it, we make it almost totally unavailable for human use; in that sense, it can be lost. More precisely, it is misplaced.

Changing lead content of the layers of snow in the Greenland Ice Sheet. A perceptible rise is seen, paralleling the rise in lead smelting in 200 years of industrial growth. A more recent, sharper rise corresponds to the introduction of lead additives in the 1920s and the greatly increased use of leaded gas since World War II.

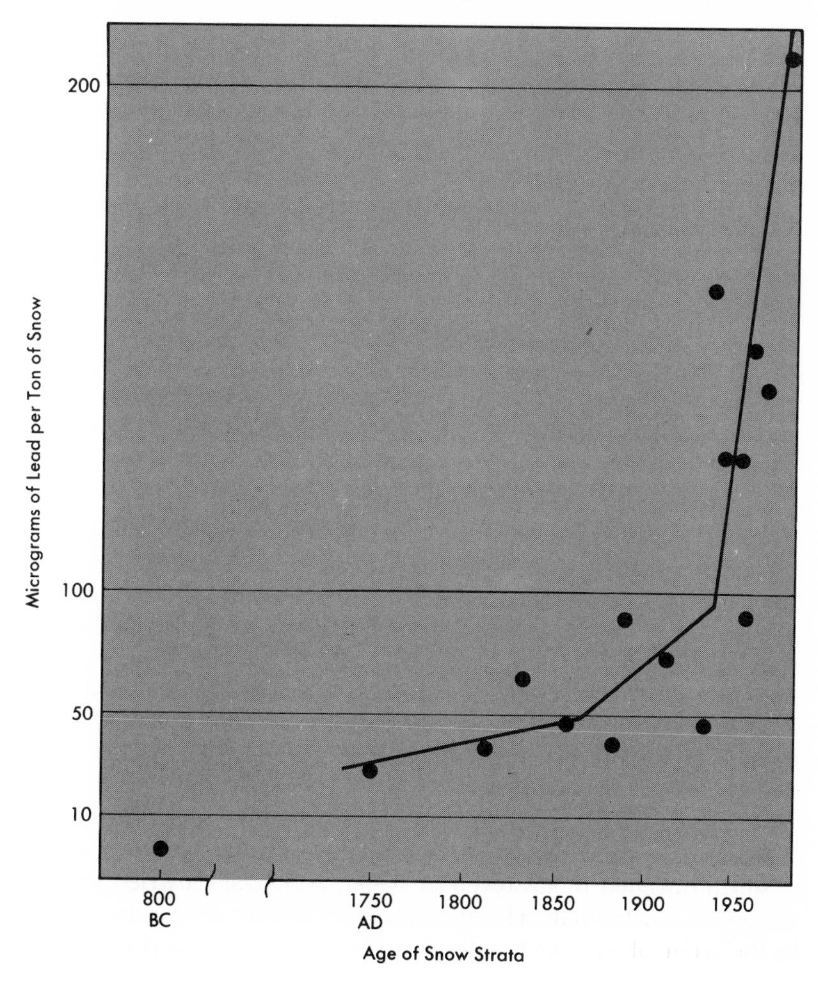

These elemental cycles do not have the strong, self-correcting mechanisms of the oxygen cycle or even of the more vulnerable carbon cycle. As long as the sulfur is within a lump of coal, in a furnace or even a stack, it can be recovered and used; once it goes out the stack it is essentially irrecoverable. The same is true for all the other elements that are going up in smoke or coming out of engine exhausts. The winds that carry them away—and we would stifle in our polluted air without the winds—ensure that the valuable components of the pollution are scattered far and wide. We have neither the competence nor the energy resources to recollect them. And the quantities thus dispersed are increasing every year.

Lead. While the quantities of most elements mobilized by man are probably much less than those mobilized in the natural sedimentary cycle, man moves lead in much larger amounts each year than does nature. The flow of lead to the oceans may be 40 times the primitive level, primarily as a result of the lead that is added to gasoline, emitted from automobile exhausts, and brought down in rain[18]. Clair C. Patterson first called attention to the fact that lead was becoming ubiquitous in our environment in the early 1960s and was rising with industrial use and automobile use[19]. Both American and European studies have found that plants and soils near highways contain more lead than those at a distance[20], but soil and plants are being contaminated by lead fallout over much wider areas as it first rises into the air, is dispersed, and then brought down in rain. No beneficial effects of lead in soil, plants, man, or other animals are known. Its climatic effects have already been mentioned. Although it is primarily as a component of high-octane gasoline that lead reaches the air, it is used in other ways as well. Every source of environmental lead deserves careful investigation, but there is little doubt that elimination of lead as a gasoline additive would reduce the lead problem to a small fraction of its present size.

Cadmium. Cadmium is another heavy metal that is appearing in air, food, and water. Like lead, it has no beneficial effects on living things and is hazardous in relatively low concentrations. Unlike lead, there is no single dominant use responsible for its appearance in the environment. There are no cadmium ores, as such, and the metal is produced as a by-product in refining other metals, primarily zinc. A certain amount of cadmium is in most multiple-metallic ore, however. Cadmium dust, fumes, and mists are common by-products during the refining of zinc, copper, and lead as well as during the extraction of cadmium[21].

An Environmental Protection Agency report estimates the total cadmium discharged to the atmosphere in 1968 at 4.6 million pounds—45 percent from the processing and refining of cadmium-bearing ores and 48 percent from the reprocessing of cadmium-containing products. Cadmium is ubiquitous because of its association with other metals and thus is a component of thousands of household and commercial products. Some

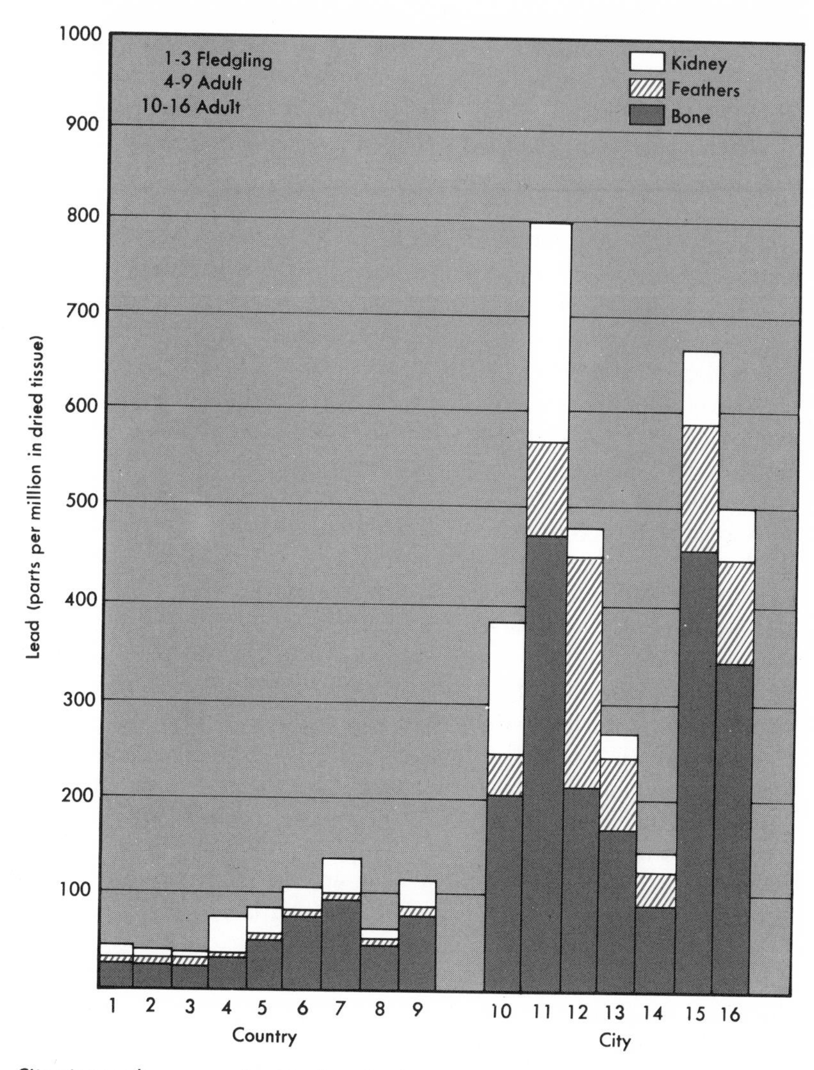

City pigeons have more lead in their feathers, bones, and kidneys than do country pigeons, as shown in this striking comparison from a Philadelphia study. Although sources of the lead may have been various, much of the lead in the bodies of the city pigeons probably came from automobile exhaust.

cadmium is lost in the manufacture of these products and more is lost in their use. For example, rubber tires contain 20 to 90 parts per million cadmium, as an impurity in zinc oxide used in rubber production, and cadmium is emitted into the air by the wear of tires during vehicular travel. At the end of the line, cadmium is emitted in the incineration of tires, plastic bottles, automobile seat covers, furniture, floor coverings, numerous other items made of polyvinyl chloride, and other cadmium-containing solid waste[22].

Mercury. Mercury, asbestos, and beryllium have been classified as hazardous air pollutants by the Environmental Protection Agency. Under the 1970 Amendments to the Clean Air Act, a "hazardous" pollutant is one to which even slight exposures may cause serious illness or death. This step underscores the fact that mercury, only recently recognized as an environmental pollutant through its use as a fungicide on seed grain and discharged to water from chloralkali plants, is present and possibly dangerous in the air as well.

Mercury is a natural component of soil and sea water and, unlike most heavy metals, it is partly volatile at ambient air temperatures. It therefore enters the air as mercury vapor from water and from soil, the latter especially over mercury deposits and areas of mineralization. Yet natural background levels or the extent to which man is adding to them, have not been established. An airborne survey by the St. Louis Committee for Environmental Information, with the technical assistance of the Center for the Biology of Natural Systems at Washington University, found mercury vapor discharge from coal-burning power plants, municipal incinerators, and industrial plants in December 1970 and January 1971 in Missouri, Illinois, and Wisconsin[23]. The source of the mercury from the power-plant stacks was undoubtedly the mercury content of the coal burned; that from the incinerators was probably mostly from paper and plastic, incorporated into these products during manufacture and released in burning.

The federal standard as proposed limits emissions to five pounds of mercury from any installation in any 24-hour period. This would have little effect on current emissions from small power plants, but would require emission controls on larger ones[24]. According to this standard, the concentration of mercury in ambient air in any one place would be held down, but a region's total output could remain high if power were produced in many small plants rather than in few large ones.

The hazard to health of mercury emissions is only part of the problem. Another question is whether volatile mercury is being continually recycled from air to water and soil and back to air. A third question is what happens to mercury fallout that directly or indirectly reaches water. Inorganic mercury compounds and the elemental forms of mercury can be transformed by microorganisms in the sediments of lakes and rivers to the more toxic alkyl mercury compounds, including methyl mercury—to people, the most hazardous of all forms of mercury. Fish can accumulate methyl mercury without apparent harm, and it can then be passed on to people who eat the fish[25]. This is one of those unpleasant surprises the environment sometimes holds in store for us when we act in ignorance.

Beryllium. The highly toxic beryllium is used in metal alloys and in rocket fuel. It is not as restricted in distribution as it was formerly thought to be. Beryllium has recently been found in the effluent of a cement plant[26]. It is an impurity in coal and therefore a constituent of industrial and power-plant emissions, but the most significant exposures are in the

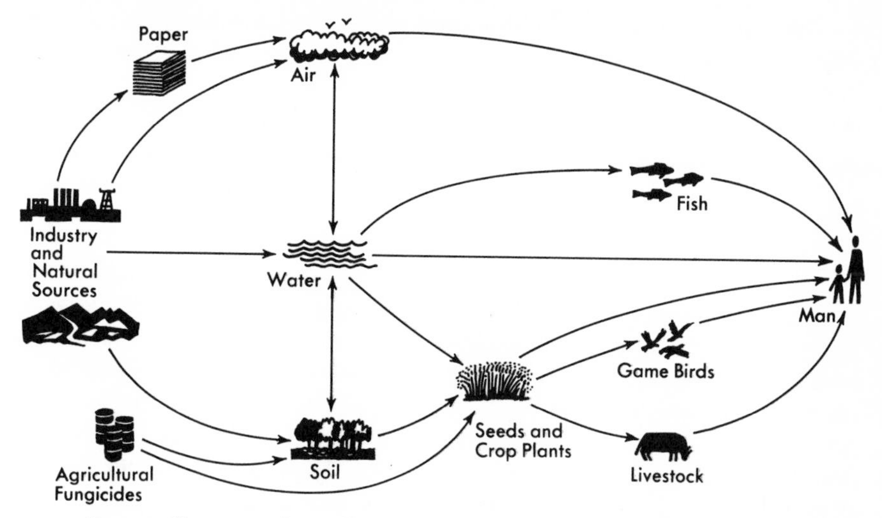

Mercury, like many other pollutants, may reach man via a number of different pathways—directly from the air, or from air to water and soil and thence through food and drink to man.

vicinity of beryllium production plants and rocket firing ranges. The former is apparently a greater hazard to vegetation, as the beryllium pollutants from these sources are soluble[27] and beryllium in this form is taken up by plants; it can inhibit their growth and can also be passed on up the food chain[28].

Asbestos. Asbestos is not an element but the fibrous form of several minerals. The quality which makes it so useful—the indestructibility of its fibers, even by fire—also assures that these fibers, although they may be reduced to very fine dust, will remain in the environment (or in the lungs of those who breathe them) for a very long time. Mining, processing, construction of buildings and ships (and the destruction of buildings, as well) are major sources. But asbestos is now incorporated into some 3000 products, many of them household items like draperies and floor tiles, and has even been put into toy cosmetics and baby talcum. It is released from these products in normal wear and tear and in waste disposal[29].

Selenium. Selenium is one of the elements essential in small doses and toxic in large doses to warm-blooded animals, including man. Cattle grazing on forage from selenium-poor soil may suffer from selenium deficiency, while vegetation growing in selenium-rich soil can be lethal to livestock. Great care needs to be taken to protect the food chains of man and other animals because the difference between a normal daily dose and a toxic dose of selenium is remarkably small. For man, the normal ingestion has been given as 0.2 milligrams per day and the toxic dose as five milligrams per day—a safety factor of only 25 in contrast to a safety factor of 4000 for chromium, another essential element.

Selenium in the air comes from fossil fuels and from the incineration of solid waste[30]. Some coals are much higher in selenium than others. The Great Lakes Research Division of the University of Michigan has found selenium in concentrations of from one to four parts per million dry weight in zooplankton from Lake Michigan. Because the soil in the Great Lakes region is selenium-poor, the source may be selenium fallout from the Chicago industrial area[31].

Nickel. Henry A. Schroeder, who has been engaged in research on trace metals for many years, considers lead, cadmium, and nickel to be the three such metals with the greatest potential toxicities[32]. The most toxic nickel compound, nickel carbonyl, is formed when hot carbon monoxide is passed over nickel. Nickel dust and hot carbon monoxide are both components of engine exhausts, coal- and oil-fired furnaces, and incinerator flues. In the case of gasoline, nickel is not only present naturally, but is sometimes used as an additive. It is not known how much, if any, of the airborne nickel is in the form of nickel carbonyl, but the possibility seems obvious and certainly deserves investigation.

Fluorine. Like selenium, fluorine is an element that is essential to man, but the particular form in which it reaches him, as well as the amount that he takes in, can make the difference between the beneficial and the harmful. Some fluoride compounds, especially hydrogen fluoride, are a serious environmental problem in the vicinity of certain industries. Fluoride is present naturally in many geological deposits in the form of fluorite, cryolite, and fluoroapatite, and whenever rock, sand, or metal is processed at high temperature, fluorides are released. They may reach the air in the vicinity of brick, glass, and ceramic factories and a variety of metallurgical industries; they come from uranium-processing plants, from coal-burning industries and power plants, and from incinerators. The major sources, however, are aluminum plants and fertilizer plants processing phosphate rocks which are high in fluoride.

Fluoride damages citrus fruit trees, pine trees, and some other plants. Plants which are less sensitive to damage may concentrate the fluoride; livestock eating contaminated forage then concentrate the fluoride in their bones and may develop fluorosis, an abnormal calcification of the bone. Livestock has been lost in the vicinity of smelters and phosphate fertilizer plants. The harmful effects of high concentrations of fluoride in particular localities have long been recognized. More recently, the possibility has arisen that fluoride fallout from industrial uses has increased to the point that it could be contaminating water, soil, and plants on more than a local scale and that it might become a widespread, low-level, long-term environmental problem. Meanwhile, many communities continue to add fluoride to their drinking water without careful evaluation of the many environmental pathways.

There has been limited sampling of fluoride in urban air, but the concentrations found have all been below the strictest standard enacted anywhere—New York's standard of 2.8 micrograms per cubic meter of air. In most cases, the concentrations were below the level of detectability [33], but air is only one source of fluoride. The fact that fluorides injure plants and accumulate in the bones of both livestock and wild mammals suggests the need for a more careful check of fluoride emissions from all sources. Plant damage from sulfur and from photochemical smog is now much more extensive than was anticipated in the past; examination of fluoride-sensitive species downwind from possible sources could provide an early warning of fluoride contamination.

Poisons, Intentional and Otherwise. The phenomenal growth of the chemical industry in the past 30 years has been based in large part on organic (carbon-containing) compounds. Living cells and dead organic matter like coal contain an immense variety of molecules in which carbon is combined with other elements in many different molecular structures. An almost unlimited number of carbon compounds can be isolated from plant

The spectacular rise in the production of synthetic organic chemicals has added a variety of new pollutants to the air. Pesticides, herbicides, and industrial chemicals are being widely dispersed, and the more persistent are contaminating many aspects of the environment.

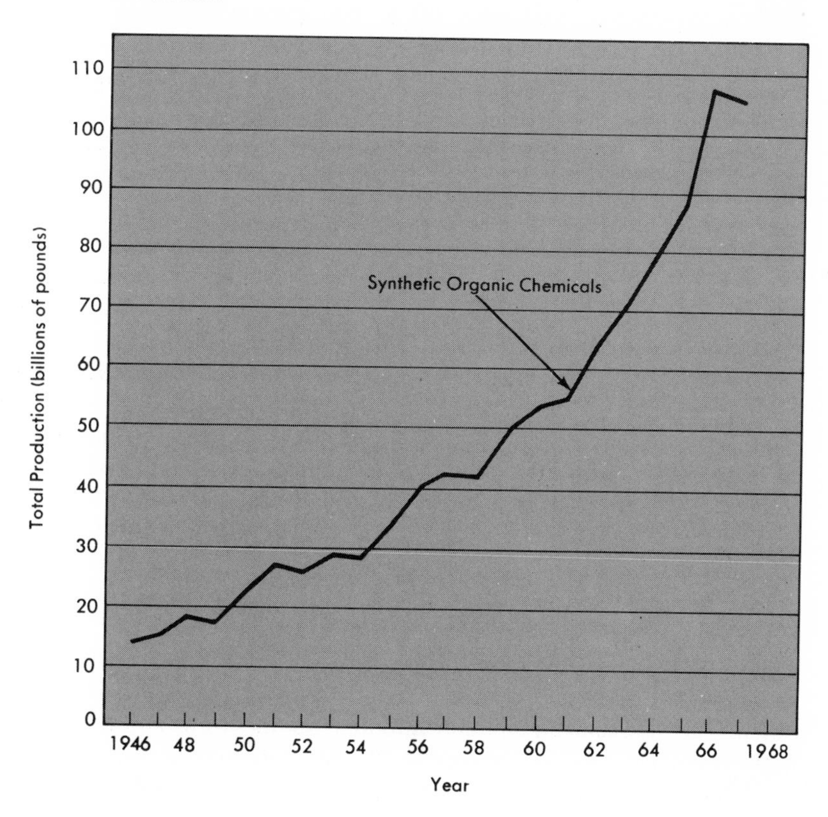

material or synthesized, forming the basis for products ranging from penicillin to synthetic rubber. While the contamination of the air by some toxic substances is an unfortunate side effect of our technology, others are manufactured for the very reason that they are poison. They are intended to interfere with nature, but they sometimes do so in unintended ways.

Insecticides and herbicides, for example, are designed to kill unwanted insects and plants. One of the great advantages of the chlorinated hydrocarbon insecticides (DDT and related compounds)—their persistence—has proved to be their greatest disadvantage as well. Nature has no known way of degrading them and therefore they remain in the biosphere for many years. Of the 126,000 tons of chlorinated hydrocarbons sold annually, more than half probably enter the atmosphere each year. Much of the spray escapes into the air without ever reaching its target, and some of what does reach plants or soil is later vaporized or is absorbed on dust particles and picked up by the wind. Moving with the wind, these hydrocarbons travel to every corner of the globe, thousands of miles from pesticide use; they come down in the rain over cities or over remote waters. These compounds persist and are accumulating in the bodies of fish and birds in the Arctic and Antarctic[34]. Effects on the reproductive systems of birds are threatening several species[35].

Organophosphate insecticides persist only for weeks or, at most, for months and are therefore not a global problem. But they are more toxic than the chlorinated hydrocarbons and so present a greater hazard in the air of the immediate neighborhood where they are sprayed. As more and more insect species have developed resistance to DDT and its relatives, the organophosphates have been used in increasing amounts.

There are also organophosphates intended to poison people, rather than insects. Should these nerve gases ever be used in war, their transport on the winds and the resulting horrors are incalculable. Very small amounts are lethal, either inhaled or absorbed through the skin. Food contaminated with them can destroy people, animals, and birds. In the spring of 1968, the country had a glimpse of their dangers when a test with VX, one of the organophosphate poisons, went awry. A slight malfunction in equipment, combined with inadequate meteorological monitoring and failure to appreciate the possible damage to grazing animals, combined to produce an accidental release killing thousands of sheep[36].

Shortly thereafter scientists on the Colorado and St. Louis Committees for Environmental Information exposed hazards in the storage and transport of chemical weapons. The Colorado Committee pointed out the danger of storage at the Rocky Mountain Arsenal, where the tanks of gas were in line with the north-south runway of Stapleton International Airport in Denver and therefore vulnerable to aircraft accidents. Both groups called congressional and public attention to the transportation hazards. In the event of an accident in transit, the gas could quickly become airborne and kill thousands of people in the path of the wind. As old chemical wea-

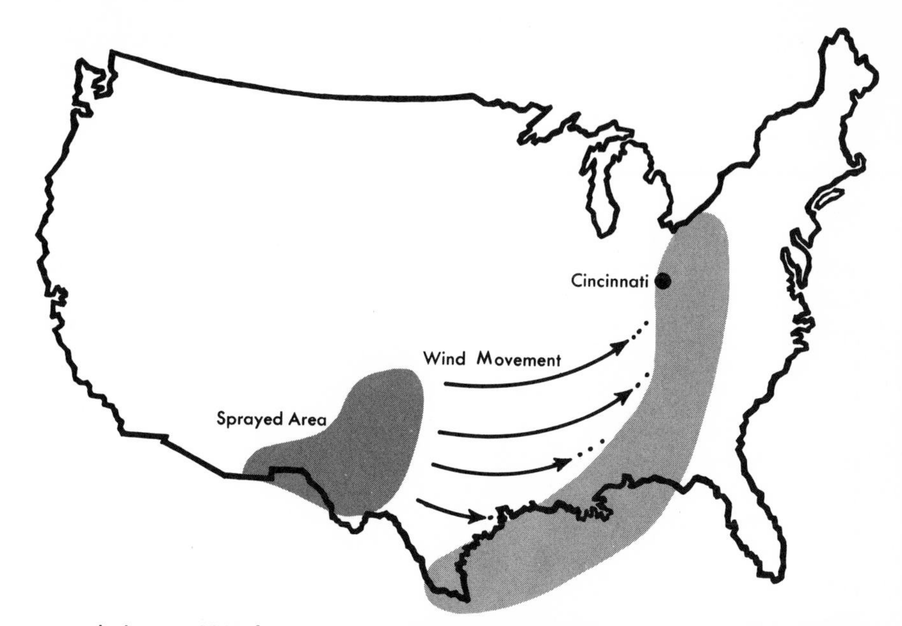

In January 1965, five to six tons of dust per square mile were deposited in Cincinnati, Ohio, by a light rain at noon. The dust had been picked up in a large storm the day before in western Texas where pesticide use is heavy and contained appreciable amounts of DDT, DDE, chlordane, and trace amounts of heptachlor epoxide and dieldrin, totaling 1.3 parts per million of chlorinated hydrocarbons. At the time the dust-fall occurred in Cincinnati, the dust cloud resulting from the storm stretched for 1500 miles in a 200-mile-wide band extending easterly from the southern tip of Texas north to Lake Erie.

pons are phased out and new ones stockpiled, the possibility of this most deadly form of air pollution will remain[37].

Bacteria intentionally disseminated in the air in warfare would be millions of times more hazardous than chemical weapons. Man has learned to produce biological weapons, to manipulate them, to disseminate them—for example, in the form of aerosols (mists of very fine particles)—but he has not learned to control them. Once released they may propagate, spread, evolve, develop relationships with other living things in ways which cannot be altogether foreseen. The destructive potential of biological weapons is in a range comparable to nuclear weapons, yet the production technologies involved are extremely cheap and theoretically very simple. Biological warfare would present a terrible danger to the attacked, a lesser, but still potent, danger to the attacker, and, in the end, a danger to the human race[38].

Herbicides, developed for use in agriculture and along highways, railroads, and waterways, are also poisons which become airborne. A field survey of air pollution injury to plants in the city of St. Louis identified herbicide injury to several species. The source appeared to be industrial plants manufacturing or handling herbicides[39]. Herbicides have been used on the largest scale by the United States in the Vietnam war to defoliate forests—especially near roads and waterways, but often over extensive areas—and to destroy food crops in order to deny the enemy food.

As almost invariably happens when material of this sort is consigned to the air, only part of it reaches its ground target. Forest insecticide spray programs in mountainous areas of the United States have found that vertical eddy motions interfere with the fallout of the drops, picking them up and carrying as much as 25 percent of the material away from the target area. By 1967, food crops and commercial plants and trees outside the sprayed areas in Vietnam (in some cases, as much as 30 miles away) were reported to have been severely damaged[40].

Although military secrecy and the difficulty of making ground studies in combat areas has limited the scope of investigations, reports by scientists have revealed that a fifth to a half of South Vietnam's mangrove forests have been completely destroyed, while between a third and a half of the trees in the hardwood forests are dead. Rubber plantations have suffered and the problems of forest recovery appear more and more serious as time goes on. The recovery of cropland needs further investigation. Effects on soil and wildlife are not clear; secondary effects, such as erosion, leaching of nutrients, and pollution of waters may seriously hamper recovery[41].

There are other chemicals which, although not intentional poisons and not intended for dissemination in the open environment, have properties similar to some of the poisons mentioned above and create similar environmental problems. For example, the polychlorinated biphenyls (PCBs) are industrial chemicals closely related to DDT. They have a wide variety of uses and, like DDT, are persistent, but somewhere in the process of manufacture, transportation, use, and disposal, they are escaping into the environment, may become airborne, and are being found in the bodies of birds and fish[42].

When synthetic fibers, plastics, and other products of the organic chemical industry are consigned to incinerators, the emissions may contain highly toxic vapors and particulates. When burned, discarded plastic housewares, textiles, packaging, and toys of polyvinyl chloride (PVC) and polyvinylidene chloride (Saran) emit hydrogen chloride—irritating in low concentrations, dangerous in high concentrations, and corrosive to materials. A recent report projects a rise in hydrogen chloride emissions from municipal incinerators from about 6500 tons nationally in 1970 to more than 100,000 tons 30 years hence. This projection is based on expected increases in population and in refuse per capita, on growing substitution of plastics for paper and natural fabric, and on an unfortunate technical problem in the control of incinerator pollution. The type of control device that will best reduce hydrogen chloride, polynuclear hydrocarbon, and nitrogen oxide emissions is a "wet scrubber," but electrostatic precipitators and fabric filters are more efficient than wet scrubbers in the removal of particulates and volatile metals. The projection assumes that the choice will be to improve particulate control at the expense of the other pollutants[43]. Even

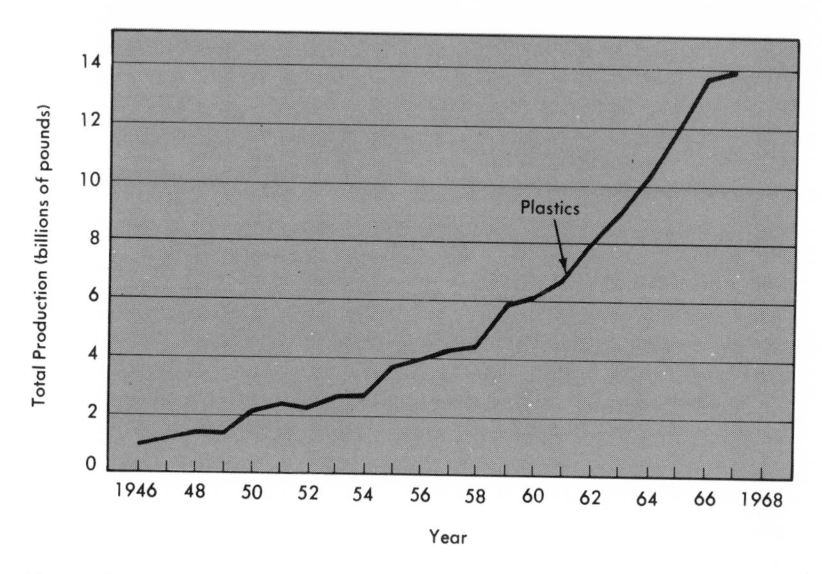

The rapid increase in the production of plastics is bringing increasing loads of plastic to municipal incinerators. When burned, some of these materials—polyvinyl chloride (PVC) and Saran—emit hydrogen chloride which is irritating in low concentrations, dangerous in high concentrations, and corrosive to materials.

if this is the technological choice, there are *social* choices which could prevent hydrogen chloride from increasing by a factor of 15: Refuse per capita could be held steady or reduced, or the substitution of plastics for paper and natural fabrics could be halted or reversed.

Other plastics which are not chlorinated, as well as inks and dyes, are probably important sources of the small amounts of polynuclear hydrocarbons appearing in incinerator effluent. Even small amounts are of concern because of the role of hydrocarbons in photochemical smog and because some of them have been found to be carcinogenic. The incinerator report just mentioned projects a rise in hydrocarbon emissions from these sources from less than 5000 tons nationally in 1970 to more than double that by the year 2000 for the same reasons that a rise in hydrogen chloride emissions is projected.

Phthalates, a class of chemicals generically related to thalidomide, are used to make common plastics soft and flexible. Sitting in a new car on a hot day may provide the most intensive exposure to the vapors of phthalate plasticizers that the ordinary citizen is likely to encounter, but so nearly ubiquitous are these materials that their effects have begun to come under question in situations varying from blood-transfusions to space probes[44].

Clean Water, Dirty Air

The many pollutants in the air from the incineration of solid waste indicates that, unless materials are reused, they are likely to become pollutants in some other part of the system. Burning of sewage sludge can pollute the air, while airborne bacteria from large sewage treatment plants—particularly when they are near centers of population—may represent a public health hazard. Investigators from the Deseret Test Center found fecal bacteria in the fine droplets carried from trickling filter bed sewage treatment facilities at a distance of 0.8 miles downwind, the greatest distance at which they took samples[45]. Cooling towers attached to power plants may cut down the thermal pollution of rivers, but increase the thermal pollution of the air.

The other side of the coin is the pollution of soil and water from efforts to clean up air pollution. Wet scrubbers may carry pollution removed from the stacks into the nearest river. Tons of particles collected from baghouses and electrostatic precipitators (the two most common particulate control devices) usually wind up in landfills, and cities are running out of space for disposing of their solid waste. *UC News* recently hailed "A simple and inexpensive solution to one of the world's most troublesome air pollution problems."[46] The problem: sulfur oxides from stack gases. The solution: clean them out with seawater and discharge the contaminated seawater back into the ocean. The unknown: the effect on marine life of the increase in sulfates and acid in the seawater[47].

Radioactive Fallout

Radioactive contamination of the atmosphere from nuclear fallout was the first example of human intervention in the environment to rouse widespread public concern. The pattern of events associated with nuclear testing has now been repeated in the cases of other kinds of contamination: *basic scientific discovery* followed by *technological application* and *immediate use,* in spite of *ignorance of biological effects;* then belated recognition of *widespread environmental distribution* and *serious biological risk.*

In the case of radioactive fallout, growing public concern led to partial control through the treaty outlawing atmospheric testing of nuclear weapons. Human exposure from weapons fallout is now estimated to be down to less than a tenth of natural background radiation[48]. However, the test-ban treaty was not signed by China or France, additional powers may acquire nuclear capability, and the treaty applies only to atmospheric testing and not to underground tests which may also contaminate the atmosphere—although the hazard is much less. As long as nuclear weapons are manufactured and stockpiled, there are hazards from accidental release at some stage in the process of mining, manufacturing, transport, and storage of both weapons and waste; installations of and training with finished weapons; and, finally, the possibility of their use in war.

Having lived with nuclear weapons for a generation without their use in any of the numerous wars that have broken out since the end of World War II, we tend to forget that nuclear weapons present the *greatest of all hazards* to the contamination of our environment via the atmosphere. We are periodically reminded when France or China carries out an atmospheric test, when an underground test by the United States or the U.S.S.R. vents radioactivity, or when international tension threatens to draw the nuclear powers into a confrontation. Understandably, when we *are* reminded, we think of the catastrophic, immediate destruction from blast and fire, followed by suffering from radiation sickness, and then the danger to the human species for generations to come. All too seldom has it been recognized that other species would be injured as well and some perhaps totally destroyed, which would have additional profound repercussions on human society and human life.

Partly as a result of tracing the radioactive products of atmospheric testing, we now know that a nuclear war would carry radioactive isotopes around the world—concentrated at first in the latitude bands where the explosions took place, then gradually spreading north and south. Some isotopes would remain in the earth-atmosphere system for thousands of years. Ecological effects of fallout on plants and animals would be profound. The destruction of urban civilization in the warring countries and possibly in neighboring countries would be a possible consequence of the combination of blast, fire, and radiation effects, while recovery of the affected life-support systems might be very slow. Many future generations of people throughout the world would be affected[49].

Early in the atomic age, the peaceful uses of nuclear explosives were presented in glowing terms: water for the thirsty desert-dweller, food for the hungry in overpopulated areas, productive land for the Latin American peasant, power for the underdeveloped nations, profits for the industrialists of developed nations, new harbors and canals for international trade, a key to heretofore unavailable oil and minerals, new knowledge for the scientist, and so on[50]. So far, every specific peaceful use of nuclear explosives that has been proposed has been slowed or halted by environmental hazards, of which the most common is the danger—and, in some cases, the certainty—that radioactive nuclides would become airborne, exposing people in the vicinity to radioactive contamination. "Vicinity" could mean a large geographical area in the case of a project like the building of a canal, with nuclear explosives used for excavation. One of the most massive projects to be suggested—and finally abandoned after a six-year study—was the nuclear digging of a new canal across the Isthmus of Panama.

Nuclear energy for the generation of electric power, however, has grown rapidly and is often described as "clean" in contrast to fossil-fuel burning. In fact, emissions from nuclear power reactors and the fuel-reprocessing plants which must accompany them are not dirty in the usual sense. But radioactive fallout differs in one central and profound way from dirt and chemical fallout. It is this difference which makes the use of nuclear explosives infeasible. And it is this difference which makes clean nuclear power a dubious choice over dirty fossil-fuel power. Radioactive pollution emits energetic particles which can destroy living tissue.

Chemical pollution is made up of stable elements which change only by combining in various ways with other stable elements, or by being transformed under certain conditions of heat and humidity from a solid to a gas phase or vice versa. Chemically identical, but highly *un*stable are the radioactive isotopes of the elements. They emit energetic rays or particles which can destroy the materials with which they come in contact, including those which make up living cells. When a radioactive isotope emits an energetic particle or ray, it may cease to be what it was and become a different element. Thus, after a nuclear weapons test, antimony 131 may appear in the atmosphere. It can emit an energetic particle and become tellurium 131, which can emit another energetic particle and become iodine 131. Each of these transformations is a step in a decay chain, with the final product being stable, nonradioactive iodine. Some of the steps in a decay chain are very rapid; some very slow. Since the radiations are not emitted at regular time intervals, the lifetime of a single radioactive atom cannot be measured. By averaging the time of these emissions for many, many atoms of the isotope, it is possible to arrive at a *half life*—the average time for half the atoms of this isotope to emit their radioactivity and decay to the next step in the decay chain. The half life of iodine 131 is about eight days, but another isotope of iodine—iodine 129, found in nuclear explosions and also appearing in the effluent of nuclear fuel reprocessing plants—has a half life of 17 million years.

Winds can carry fallout great distances and rain can wash it out of the atmosphere rapidly. It can then be taken up by grazing cattle, whose milk will be contaminated, or can contaminate vegetables and cereals. Thus, even short-lived radionuclides like iodine 131 can represent a hazard, as well as the long-lived strontium 90 (half life about 28 years), cesium 137 (half life about 27 years), and others. The short and simple grass-cow-milk-human food chain is only one of many which may become contaminated by radioactivity. In some cases, the radioactivity is increasingly concentrated as it moves up the food chain. Other chains are long and complex, and include geographical dispersion through migratory insects and birds. Some unusual food chains have produced unpleasant surprises in this respect. For example, lichens get their food directly from the air, and therefore the lichen-caribou-human food chains of the far North became heavily contaminated, although weapons fallout was much lighter in the northern latitudes than it was farther south. Some species are more sensitive to radiation than others. Cockroaches and crabgrass can survive radiation doses that would kill honeybees and pine trees.

Atmospheric contamination from nuclear power reactors and fuel reprocessing plants has been primarily in the release of radioactive krypton and zenon. Like their stable counterparts, these gases are chemically inert and therefore present less hazard per unit of radioactivity than radionuclides which become incorporated in human bone (strontium 90) or thyroid tissue (iodine 129 and 131). However, krypton 85 is emitted in very

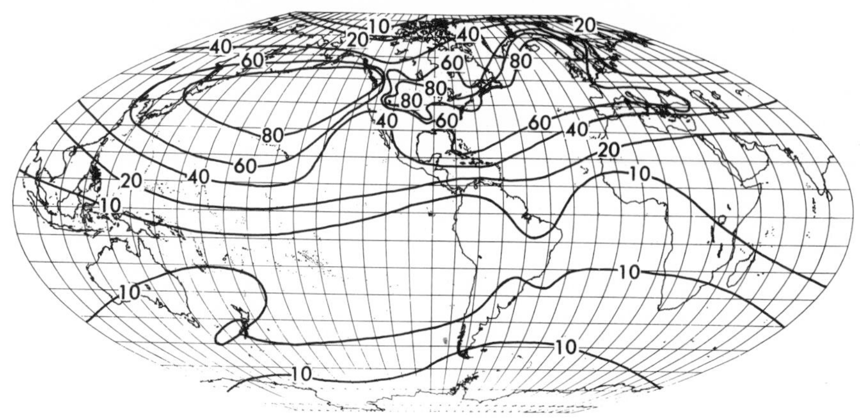

Strontium 90 in soil, in millicuries (thousandths of a curie) per square mile, mapped after a survey in 1965 to 1967. Little was known about the rate of fallout or the movement of radioactive isotopes through food chains when atmospheric nuclear testing began. Fallout proved to be more rapid and less uniform than expected. "Hot spots" far from the test sites were found, and certain food chains became heavily contaminated where fallout was comparatively light. The maximum fallout occurred in 1966, with the Northern Hemisphere most heavily contaminated. There has been little change in the strontium 90 content of soil since 1967 because the radioactive decay of the old deposits is being balanced by the fresh fallout from Chinese and French tests.

large quantities, so that a growth in the nuclear generation of power can increase radioactive krypton in the atmosphere considerably before its decay rate and other removal processes take as much out of the atmosphere in a given year as is being added. When radioactive krypton is inhaled, it can be carried throughout the body in a dissolved state, giving off radiation which may injure body cells[51]. Some iodine 131 is emitted to the air, but iodine 129, whose significance was largely overlooked in weapons testing fallout, may prove to be a more serious hazard. Because of its extraordinarily long half life, iodine 129 can be considered essentially a permanent contaminant constantly building up in the environment[52].

So much for routine operation. If this were the only problem with reactors, it would not be so difficult to balance their environmental cost against that of fossil-fueled plants. The great imponderable is the danger of an accident in which radiation is released to the environment. Such an accident could be external, from an earthquake or other natural or man-made disaster, or internal, from the malfunctioning of the reactor and the failure of its safety devices. Because today's reactors will not explode like a bomb, blast, heat, and intense initial radiation are not to be feared, but if their containment is breached, they could release quantities of radionuclides comparable to the fallout that follows a nuclear explosion. The area—and people—downwind from such an accident would get an immediate dose of very high levels of radiation, and its effects could be extensive both in space and time. Accidents do happen, although so far there has not been an accident of this nature in a commercial power reactor and many safety measures are required in both construction and operation. However, in 1970, there were only 15 power reactors functioning in this country, the oldest one only 17 years old. Within the next few years, more than 80 plants which are now being built or which have been proposed are expected to be in operation. Each new reactor adds to the risk—a risk that cannot be measured, but is not negligible—that an accident will happen[53].

. . . and Other Accidents. Although reactor accidents present the most dramatic danger, other accidental releases of hazardous volatile products occur. In Poza Rica, Mexico, in 1950, a plant recovering sulfur from natural gas released hydrogen sulfide to the air. Malfunction of equipment, fog, weak winds, and a low inversion proved to be a lethal combination. Three hundred were sickened and 22 died[54].

Effects on Ecosystems

Investigations of the effects of one or more air pollutants on one or more species of plant or animal life are important; evaluations of the damage to crops and forests in dollars can help us understand some of the economic costs of air pollution. Neither approach is sufficiently broad or basic to give us an insight into the real dimensions of pollution's environmental effects.

In an article in *Science*[55], ecologist George Woodwell pointed out that exposure of the environment to high doses of radioactivity (as in the well-known Brookhaven experiments), to repeated fires, or to severe air or water pollution has similar effects on the structure and physiology of ecosystems. In terrestrial plant communities these effects are: a shift away from mixed oak-pine forests with the elimination of pine; with more severe pollution, a shift away from oak forests toward shrubs; with still more severe pollution, a shift from shrubs toward sedge; and, finally, in the most severely polluted areas, no survival of higher plants—only mosses and lichens. The destruction of a forest community will be accompanied by the loss of nutrients normally recycled and by the loss of herbivorous species whose food and habitat have disappeared.

Woodwell sums up the significance of what is happening to ecosystems attacked by pollutants:

> Biological evolution has divided the resources of any site among a large variety of users—species—which, taken together, confer on that site the properties of a closely integrated system capable of conserving a diversity of life. The system has structure; its populations exist with certain definable, quantitative relationships to one another; it fixes energy and releases it at a measurable rate; and it contains an inventory of nutrients that is accumulated and recir-

culated, not lost. The system is far from static; it is subject, on a time scale very long compared with a human lifespan, to a continuing augmentive change through evolution; on a shorter time scale, it is subject to succession toward a more stable state after any disturbance. The successional patterns are themselves a product of the evolution of life, providing for systemic recovery from any acute disturbance. Without a detailed discussion of the theory of ecology, one can say that biological evolution, following a pattern approximating that outlined above, has built the earth's ecosystems, and that these systems have been the dominant influence on the earth throughout the span of human existence. The structure of these systems is now being changed all over the world. We know enough about the structure and function of these systems to predict the broad outline of the effects of pollution on both land and water. We know that as far as our interests in the next decades are concerned, pollution operates on the time scale of succession, not of evolution, and we cannot look to evolution to cure this set of problems. The loss of structure involves a shift away from complex arrangements of specialized species toward the generalists; away from forest, toward hardy shrubs and herbs; away from those phytoplankton of the open ocean that Wurster proved so very sensitive to DDT, toward those algae of the sewage plants that are unaffected by almost everything including DDT and most fish; away from diversity in birds, plants, and fish toward monotony; away from tight nutrient cycles toward very loose ones with terrestrial systems becoming depleted and with aquatic systems becoming overloaded; away from stability toward instability, especially with regard to sizes of populations of small, rapidly reproducing organisms such as insects and rodents that compete with man; away from a world that runs itself through a self-augmentive, slowly moving evolution, to one that requires constant tinkering to patch it up, a tinkering that is malignant in that each act of repair generates a need for further repairs to avert problems generated at compound interest[55].

When we realize that to these unwitting effects of man on his environment are added the gobbling up of natural ecosystems by cities and suburbs and the intentional change from natural diversity to the simple, unstable ecosystems of our monoculture-dominated agriculture, Woodwell's final question is a sobering one: What fraction of the earth's biota is needed to sustain the earth as we know it?

In the light of the known and estimated effects of air pollutants on atmospheric processes and biological systems and the vast gaps in our knowledge, it is hard to understand how the dispersal of pollutants can still be touted as a method of air pollution control. Yet "meteorological management"—that is, the timing of emissions to take advantage of favorable wind conditions, and high stacks to disperse smoke plumes—is still seriously advocated as an answer to some pollution problems, and dispersion models are considered an important tool in determining what quantity of pollution to permit in a given city. Dispersal is only a temporary and partial solution to the urban concentration of pollutants. It is this urban concentration that has attracted most attention and concern. Although far from the whole problem, it is certainly of vital importance because of its direct effects on the quality of urban living and its hazard to human health.

[1] A. Clyde Hill, "Vegetation: A Sink for Atmospheric Pollutants," *Journal of the Air Pollution Control Association,* June 1971, 21:341–46.

[2] Fritz Went, "Plants Monitor Pollution," *Scientist and Citizen* (now *Environment*), October 1965, Vol. 8, No. 1, 6–9.

[3] A series of papers by R. W. Stark, P. R. Miller, F. W. Cobb, Jr., D. L. Wood, J. R. Parmeter, and E. Zavarin in *Hilgardia,* the journal of the California Agricultural Experiment Station, May 1968, Vol. 39, No. 6.

[4] The *San Francisco Chronicle,* January 17, 1969.

[5] C. S. Brandt and Walter W. Heck, "Effects of Air Pollution on Vegetation," in A. C. Stern (ed.), *Air Pollution,* Vol. I (New York: Academic Press, Inc., 1968), pp. 401–43.

[6] W. W. Kellogg et al., "The Sulfur Cycle," *Science,* February 11, 1972, 175: 595.

[7] *Oil and Gas Journal,* November 27, 1967, 71–74.

[8] *Chemical Week,* April 11, 1968. Data from a paper presented by S. Katell and H. Perry at the Society of Mining Engineers meeting in Las Vegas, September 1967.

[9] H. A. Menser and H. E. Heggestad, "Ozone and Sulfur Dioxide Synergism: Injury to Tobacco Plants," *Science,* July 22, 1966, 153:424–25.

[10] Samuel N. Linzon, "Economic Effects of Sulphur Dioxide on Forest Growth," paper presented at the 63rd Annual Meeting of the Air Pollution Control Association, St. Louis, Missouri, June 1970.

[11] Clarence C. Gordon, University of Montana, unpublished data.

[12] B. Ottar, Norwegian Institute for Air Research, personal communication. The material on acidity in this area which follows is also from Dr. Ottar.

[13] Gordon, unpublished data (see [11]).

[14] D. W. Fisher, A. W. Gambell, G. E. Likens, and F. H. Bormann, "Atmospheric Contributions to Water Quality of Streams in the Hubbard Brook Experimental Forest, New Hampshire," *Water Resources Research,* October 1968, 4:1115–26.

[15] C. C. Delwich, "The Nitrogen Cycle," *Scientific American,* September 1970, 223:146.

[16] Barry Commoner, "Nature Unbalanced: How Man Interferes with the Nitrogen Cycle," *Scientist and Citizen* (now *Environment*), January–February 1968, Vol. 10, No. 1, 9–19.

[17] K. K. Bertine and Edward D. Goldberg, "Fossil Fuel Combustion and the Major Sedimentary Cycle," *Science,* July 16, 1971, 173:233–35.

[18] Paul P. Craig and Edward Berlin, "The Air of Poverty," *Environment,* June 1971, Vol. 13, No. 5, 2–9, quoting T. J. Chow and Clair C. Patterson.

[19] Clair C. Patterson, with Joseph P. Salvia, "Lead in the Modern Environment," *Environment,* April 1968, 10:66–79.

[20] For example, A. Kloke and H-O Keh, "Pollution of Cultivated Plants with Lead from Auto Exhaust," *Proceedings of the First European Congress on the Influence of Air Pollution on Plants and Animals* (Wageningen, The Netherlands: H. Veenman & N. V. Zonen, 1969). Also A. L. Page and T. J. Ganje, "Accumulation of Lead in Soils from Regions of High and Low Motor Vehicle Traffic Density," *Environmental Science and Technology,* February 1970, 4:140–42.

[21] Julian McCaull, "Building a Shorter Life," *Environment,* September 1971, Vol. 13, No. 7, 3–38.

[22] *National Inventory of Sources and Emissions—Cadmium, Nickel and Asbestos, 1968.* Cadmium, Section I, W. E. Davis & Associates, Environmental Protection Agency, Durham, North Carolina, February 1970.

[23] *Environment* staff, "Mercury in the Air," *Environment,* May 1971, Vol. 13, No. 4, 24–33. Other articles in the same issue deal with "Mercury in Man" and "Mercury in the Environment."

[24] Steven Fuller, personal communication. Mr. Fuller was a member of the mercury survey team mentioned on page 94. Coal burned in Midwest power plants has about 0.2 parts per million mercury. On the basis of the amount of coal burned, some of the new, large plants probably emit about 3000 pounds per year or an average of about ten pounds per day.

[25] A. Jernelov, "Conversion of Mercury Compounds," in *Chemical Fallout,* M. W. Miller and G. Berg (eds.) (Fort Lauderdale: Charles C Thomas, 1969).

[26] Harry Rhoades, testimony before the Senate Subcommittee on Air and Water Pollution, St. Louis hearings, October 1969.

[27] Herbert E. Stokinger and D. L. Coffin, "Biologic Effects of Air Pollutants," in Stern, *Air Pollution,* Vol. I, p. 497 (see [5]).

[28] E. M. Romney and J. D. Childress, "Effects of Beryllium in Plants and Soil," *Soil Science,* 1965, 100:210–17.

[29] I. J. Selikoff, "Asbestos," *Environment,* March 1969, Vol. 11, No. 2, 3–7.

[30] Henry Johnson, "Determination of Selenium in Solid Waste," *Environmental Science and Technology,* October 1970, 4:850.

[31] Richard Copeland, "Selenium and Mercury, Distribution in Lake Michigan Invertebrates," *Geological Society of America Abstracts,* 1970, Vol. 2, No. 7, 526.

[32] Henry A. Schroeder, "Metals in the Air," *Environment,* October 1971, Vol. 13, No. 8, 18–31.

[33] R. J. Thompson, T. B. McMullen, and G. B. Morgan, "Fluoride Concentrations in the Ambient Air," *Journal of the Air Pollution Control Association,* August 1971, 21:484–87.

[34] Justin Frost, "Earth, Air, Water," *Environment,* July–August 1969, Vol. 11, No. 6, 14–33.

[35] Kevin P. Shea, "Unwanted Harvest," *Environment,* September 1969, Vol. 11, No. 7, 12–16.

[36] Virginia Brodine, Albert Pallmann, and Peter P. Gaspar, "The Wind from Dugway," *Environment,* January–February 1969, Vol. 11, No. 1, 2–9 and 40–45.

[37] Virginia Brodine, "The Secret Weapons," *Environment,* June 1969, Vol. 11, No. 5, 12–26.

[38] Milton Leitenberg, "Biological Weapons," *Scientist and Citizen* (now *Environment*), August–September 1967, Vol. 9, No. 7, 153–67.

[39] F. O. Lanphear, "Air Pollution Injury to Plants in St. Louis," paper presented at the 63rd Annual Meeting of the Air Pollution Control Association, St. Louis, Missouri, June 1970.

[40] Arthur W. Galston, "Changing the Environment," *Scientist and Citizen* (now *Environment*), August–September 1967, Vol. 9, No. 7, 122–29.

[41] F. H. Tschirley, "Defoliation in Vietnam," *Science,* February 21, 1969, 163:779; E. W. Pfeiffer and G. H. Orians, "Mission to Vietnam," *Scientific Research,* June 9, 1969 (reprinted in *Environmental Effects of Weapons Technology,* Scientists' Institute for Public Information Workbook, New York, 1970); Matthew Meselson et al., "Preliminary Report of Herbicide Assessment Commission," paper presented at the December 1970 Meeting of the American Association for the Advancement of Science, Chicago, Illinois. For a review of this unpublished report, see Terri Aaronson, "Defoliation in Vietnam," *Environment,* March 1971, Vol. 13, No. 2, 34–43.

[42] Robert Risebrough, with Virginia Brodine, "More Letters in the Wind," *Environment,* January–February 1970, Vol. 12, No. 1, 16–27. For a more recent, complete and technical report, see *Polychlorinated Biphenyls and the Environment,* Interdepartmental Task Force on PCBs, (Washington, D.C.: distributed by the National Technical Information Service, U.S. Department of Commerce, May 1972).

[43] Walter R. Neissen et al., *Systems Study of Air Pollution from Municipal Incinerators,* prepared by Arthur D. Little, Inc., for the Division of Process Control Engineering, National Air Pollution Control Administration, U.S. Department of Health, Education and Welfare, March 1970, p. VI–31, Figure VI–7, "Total Annual Stack Emission Estimates for U.S. Municipal Incinerator Systems Reflecting Rapid Implementation of Advanced Incineration Concepts."

[44] Kevin P. Shea, "The New-Car Smell," *Environment,* October 1971, Vol. 13, No. 8, 2–9.

[45] A. Paul Adams and J. Clitton Spendlove, "Coliform Aerosols Emitted by Sewage Treatment Plants," *Science,* September 8, 1970, 169:1218–20.

[46] Don Koue, "Smog . . . Seawater Can Help," *UC News,* September 8, 1970, Vol. 46, No. 1, 1.

[47] Additional problems of water pollution from fallout of air pollutants are discussed by Julian McCaull and Janet Crossland in *Water Pollution,* another book soon to be published in this *Environmental Issues* series.

[48] Dean E. Abrahamson, "Environmental Cost of Electric Power," Scientists' Institute for Public Information, New York, 1970, p. 13. For a good summary of test fallout, see the Test-Ban Treaty Anniversary issue of *Scientist and Citizen* (now *Environment*), September–October 1964, Vol. 6, Nos. 9–10.

[49] Barry Commoner, *Science and Survival* (New York: Viking Press, 1966), pp. 64–89. See also a three-issue *Scientist and Citizen* (now *Environment*) series in which the question was debated by two groups of scientists: May–June 1965, August 1965, and February–March 1966.

[50] Ralph Sanders, *Project Plowshare* (Washington, D.C.: Public Affairs Press, 1962).

[51] Malcolm Peterson, "Krypton 85—Nuclear Air Pollutant," *Scientist and Citizen* (now *Environment*), March 1967, Vol. 9, No. 3, 54–56.

[52] John J. Russell and Paul B. Harn, "Public Health Aspects of Iodine 129 from the Nuclear Power Industry," *Radiological Health Data and Reports,* April 1971, 12:189–94.

[53] The reactor program is examined in Sheldon Novick's *The Careless Atom* (Boston: Houghton Mifflin Company, 1969).

[54] J. R. Goldsmith, "Effects of Air Pollution on Human Health," in Stern, *Air Pollution,* Vol. 1, p. 557 (see [5]).

[55] George M. Woodwell, "Effects of Pollution on the Structure and Physiology of Ecosystems," *Science,* April 24, 1970, 168:432. Copyright © 1970 by the American Association for the Advancement of Science. Reprinted by permission.

The Burdened Human

The basis of present primary air quality standards is the direct effect of air pollutants on human health. Just as present control systems seek to use meteorological mechanisms to the limit, the approach to health effects has been to use the body's defenses to the limit. Health effects are represented in air quality standards only to the extent that a specific concentration of a single pollutant can be demonstrated to cause a harmful effect on health. The standard is then set *below* this concentration to provide a margin of safety. The lowest concentrations of the major pollutants which have been found to cause harmful effects on health are listed in the appendix; the studies from which these data come are referenced, and the margin of safety for each pollutant calculated.

We have neither the monitoring information nor the medical information needed to arrive at the kind of proof required by this one-to-one, cause-effect model, and it is too simple to be well-adapted to the human body—a complex system in delicate balance. Over the course of evolution, the human species has developed physical and biochemical processes (not all of them completely understood) for maintaining this balance and protecting the body against environmental hazards—extremes of temperature, infections, injuries, imbalances and inadequacies in diet, inhaled or ingested foreign matter. Living, breathing, constantly changing bodies, each with its own hereditary and environmental history, are unlikely to respond to an environmental challenge in the simple fashion needed to fit a simple, cause-effect model.

Although it is universally recognized that a single pollutant does not ap-

pear in the air alone, this model is unable to take into account the varied effects on people as they are exposed to different combinations of pollutants, or the synergistic effect of pollutants, or the changing effects of pollutants with changing weather, or the reinforcement of an airborne pollutant's effects when the same material reaches people via other environmental pathways. The present approach also focuses on the more obvious, acute effects most likely to be produced by exposure to high doses of a pollutant, rather than the long-term effects of low-level doses which are difficult to detect and may not show up until after years of exposure.

The difficulty in proving a cause-effect relationship opens the standards to attack from industrial opponents of control who take the position that weaknesses in a proof require the weakening of a standard. Rushing to the defense of the standards, and of the monitoring and medical information on which they are based, may be fighting on the wrong battleground with the wrong weapons. It distracts us from the real issue: Are these standards adequate to protect the public health? In order to answer this question, we must look carefully at the kind of information on which they are based.

Monitoring Information

In order to relate ambient air concentrations of pollutants to health effects, the pollutants must be measurable, and measurements in one place must be comparable to those in another. Instruments for monitoring sulfur dioxide have not been standardized in the past and have been subject to interference from other substances in the air. Sulfur dioxide can change in moist air to sulfuric acid mist, for example, through interaction with iron, manganese, and vanadium particles, and thence to sulfate particles through interaction with ammonia. Sulfuric acid mist is more harmful than sulfur dioxide, and sulfur dioxide is more harmful in combination with irritant sulfate particles (especially zinc ammonium sulfate)[1]. The measures of sulfur dioxide are therefore rather crude representations of the nature and dimensions of the sulfur hazard, especially since the effect is increased by carbon particles which adsorb sulfur dioxide and carry it into the lung in greater concentrations than ambient levels would indicate.

Modifications of the monitoring technique have been worked out which minimize the interferences from other pollutants, and the EPA has now prescribed certain types of instrumentation so that measurements from different places will be more comparable, but there are still three acceptable methods of detecting sulfur dioxide. Placed side by side, they do not always show the same results[2].

A wide variety of different kinds of particles are included in measurements of suspended particulates, so that 75 micrograms of particulates per cubic meter of air (the federal standard) in the steel town of Gary, Indiana, may be considerably different from the same concentration in the commercial center of Indianapolis. The high volume sampler, the standard measuring device for these particles, is an efficient collector of suspended particles down to about 0.5 micron (a micron is one thousandth of a millimeter). It gets some of the smaller particles, particularly in heavily polluted air after the filter begins to clog, but little of the material below 0.3 micron in size. It is now well established that submicron particles may penetrate deep into the lung, and that the irritant particles produce increasing effects with decreasing size. We may therefore deceive ourselves about the extent of particle clean-up if this clean-up affects only, or mainly, the larger particles. Enforcement officers often use grayness or blackness of stack plumes to detect violations of particulate emission standards, but the fine particles are not necessarily detected this way—they may be invisible to the naked eye. The usual method of monitoring particulates at continuous monitoring stations is according to a "soiling index" or coefficient of haze (coh). This measures the dirtiness of the air by comparing the amount of light transmitted through filter paper soiled by the polluted air with the amount of light passed through clean filter paper. It is less reliable than the high volume samplers, and the EPA specifies that it should only supplement, not replace, the latter. The hydrocarbon standard is not based on the direct effect of hydrocarbons on human health, but rather on the role of nonmethane hydrocarbons in the production of photochemical smog. In order to ascertain the nonmethane hydrocarbon concentration, both methane and total hydrocarbons are measured and the methane measurement subtracted. The measurements are so inexact that methane alone is sometimes reported as greater than the total.

If there are too few monitoring stations or if they are poorly located, it may not be clear how many of the people in a health study were actually exposed to the "measured" concentration of the pollutant. A study of the sampling station and time requirements for monitoring that was done in Nashville found that, in order to get accurate measurements of the *daily* exposure of the community to sulfur dioxide, 245 stations or four per square mile would be needed, and to measure particulates (by the coh method) 120 stations, about two per square mile[3]. No city has anything approaching this density of stations, which would be fantastically expensive and would generate quantities of data, requiring a tremendous staff for maintenance and processing and leaving little money or personnel for investigation and enforcement. Fewer stations would be needed to get reasonably accurate seasonal and annual averages. But even if only annual averages were required, the concentration of sulfur dioxide presumably responsible for an observed effect on the health of a community the size of Nashville might be off by 40 percent or more if the city has only two stations. Thus, a city with one or two stations could be reporting "safe" levels of the pollutant on an annual average although these levels might be exceeded outside the immediate range of the monitoring stations. Since the margin of safety built into the annual average sulfur dioxide standard is only 30 percent, an uncertainty of 40 percent can eat up that entire safety margin.

Increased deaths during episodes of high pollution show that monitoring adequate in relation to the annual average standard is not necessarily sufficient in relation to the daily average, and there are now two sulfur dioxide standards—one for the annual average and one for the maximum allowable in any one day. To monitor for violations of the daily standard, either a dense network of stations is needed or fewer stations must be carefully located in the areas of highest pollution.

Medical Information

Air pollution has been found to be associated with increased incidence of respiratory disease, with increased symptoms in people already suffering from respiratory disease, with increased incidence of cancer, and with higher mortality rates. But disease or death is associated with *air pollution as a generalized environmental phenomenon,* rather than with any single pollutant. It is impossible to conduct a study in which one population is exposed to ambient air containing a given concentration of sulfur dioxide (along with other pollutants) while a control population breathes air of the same composition *except* for a different sulfur dioxide level, and thereby to learn the effect of sulfur dioxide alone. It is impossible to expose the same population at different times to air pollution that is the same except for a difference in particulates. In life, ambient air almost never comes in such packages. An unusual exception is air in three different parts of Chattanooga which has been found to be quite similar for sulfur oxides and particulates, but strikingly different for nitrogen dioxide. The latter was highest near a TNT plant and diminished with distance from the plant[4].

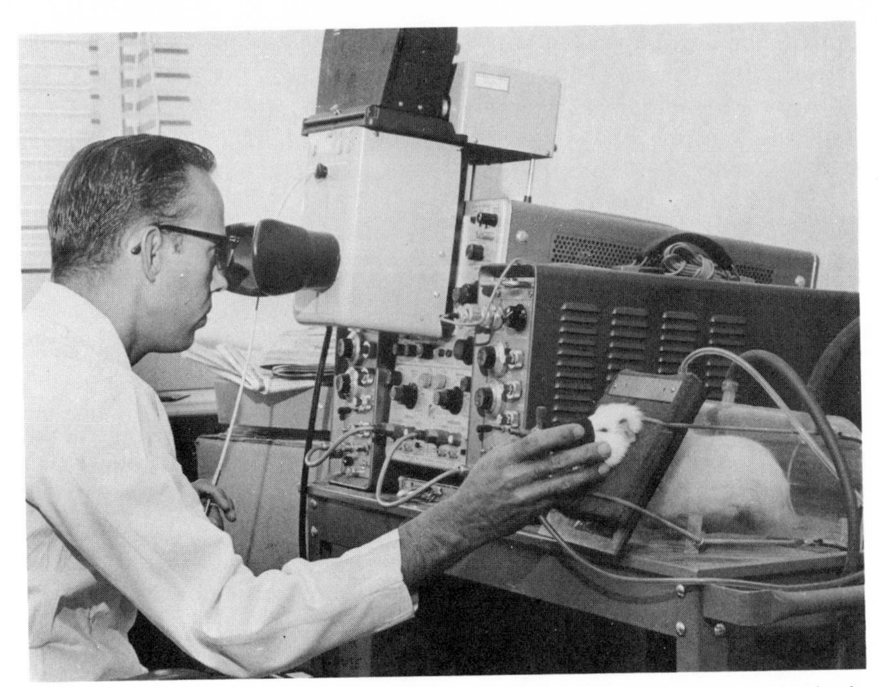

In most laboratory experiments, laboratory animals are exposed to contrived substitutes for ambient air. It is thus difficult to relate the physiologic reactions of the experimental animals to possible human reactions to exposures to ambient air. In an attempt to bypass this problem, the rate of respiration of a guinea pig who has been exposed to *ambient* air is being tested. This is part of a long-term study at The Statewide Air Pollution Research Center at the University of California, Riverside, in which control animals breathe filtered air while other groups of animals are exposed to ambient air containing the same pollution people breathe.

But if the outdoor air in a city is always a complex mixture of known and unknown pollutants, the same cannot be said of the indoor air in certain mines, factories, or other places where working with certain materials results in air pollution with a special character. It would seem that much might be learned from the field of occupational health which could be a guide in establishing the kind of single-pollutant, dose-response relationship we seek. Unfortunately, monitoring and epidemiologic studies in work places have been woefully limited, and controls often have lagged a generation or more behind the introduction of the hazardous material.

One reason for the paucity of good occupational health data is the reluctance of many industries to cooperate in long-term studies or, in some cases, even to permit health investigators on their premises. Physicians in the employ of industries have made few notable contributions, a fact that can perhaps be explained by the following sentence from a textbook for this branch of medicine:

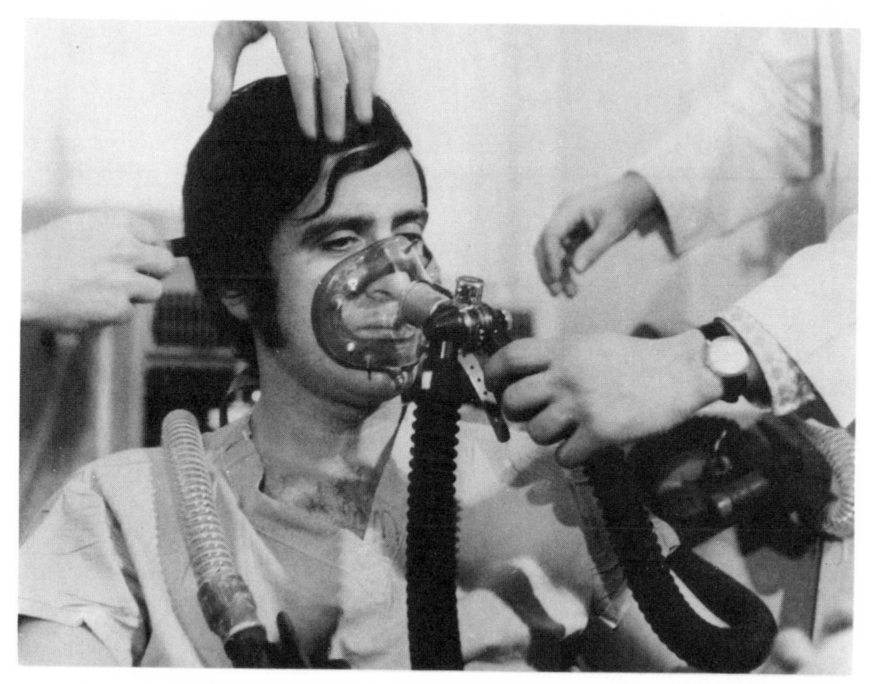

In a second attempt, the reactions of a human subject to exposure to measured concentrations of *carbon monoxide* are being observed at the Environmental Protection Agency's federal laboratory.

> The physician's place in the industrial system is quite different from that to which he has become accustomed in private practice. . . . His services are strictly ancillary to the main purpose of the business: production at a profit[5].

Occupational standards are intended to apply to a healthy, adult group exposed only during working hours. In theory, frequent medical examinations are possible and a worker suffering from in-plant exposure can be removed to a different environment. (In life, the former requirement is honored more in the breach than in the observance, while jobs in a cleaner environment are often not available in the same plant. Jobs in another industry may also be beyond the reach of individual workers, especially if they are middle-aged or live in a one-industry mining or mill town.)

In the laboratory, it is possible to control both exposures and subjects, but here the subjects must almost always be laboratory animals. There have been some experiments with human volunteers, but these suffer from the limited number of subjects and the limited duration of exposures. In addition, just *because* of the controls in the laboratory, results are difficult to apply to ambient air. Because a guinea pig can breathe pure sulfur dioxide of a given concentration under controlled temperature and humidity without showing any ill effects, we cannot assume that a similar exposure is safe for people who never breathe pure sulfur dioxide, but always inhale it in combination with other sulfur compounds, with irritant particles, with other gases, at a fluctuating rate under variable weather conditions. On the other hand, although the induction of lung cancer in a laboratory animal by exposure to certain components of ambient air may strongly suggest that the same materials may be carcinogenic in man, such a suggestion is difficult to translate into a standard permitting any particular amount of such material in the air.

Experiments housing animals in such a way as to expose them continuously to actual polluted Los Angeles air, rather than to a contrived simulation, have succeeded in eliminating some of the problems of previous laboratory experiments, although the difficulties of relating the results to humans remains. Mice, rabbits, and guinea pigs were all used, and observations continued over two and a half years. A number of biological differences were noted in the animals exposed to polluted air when they were compared to the control animals kept in filtered air[6].

The fact is that these three avenues of investigation—community epidemiology directly related to exposures to ambient air, occupational epidemiology, and laboratory research—have produced an impressive body of information about the deleterious physiological effects of air pollution. We are simply asking the wrong question when we try to draw from this body of information a series of numbers describing safe concentrations of separate pollutants. It is not that these standards are useless; they are the only "starters" we now have for getting machinery in motion to reduce air pollution which is quite clearly a hazard to human health. But the standards represent a simplistic and incomplete answer to a complex problem.

The complexity of the problem and the inadequacy of the present solution can be better understood by taking a brief look at some of the body's defenses against inhaled pollutants and some of the ways in which these defenses can be bypassed or weakened.

Breaching the Defenses

The body's defense is a marvelous mechanism with a whole series of interacting processes for maintaining, within rather narrow limits, a constant chemical and physical constitution of the internal environment and protecting the body from invasion. If it were not, few of us could survive the combinations of noxious gases, infective bacteria, and irritant particles we breathe. But although we are protected, our defenses can be bypassed, weakened, or overcome at each stage. We do not yet understand all the workings of the system when it deals successfully with inhaled pollutants; our understanding of its failures is still less complete. Nor can any of us usually perceive a breach of his own body's defenses until long after it has occurred, or confidently link it with air pollution. Nevertheless, clean air is not only more enjoyable to breathe, it is vastly better for us than polluted air.

The Respiratory Tract. The nose is the first line of defense. The hairs in the front of the nose block large particles. Then the air currents are broken into small streams by the spiral, spongy bones in the nasal passage so that turbulence occurs and those particles that gain entrance are forced against the sticky walls of the passage, with their tiny projecting hairs or cilia. The cilia move constantly, carrying mucus with some particles trapped in it toward the back of the throat where it is swallowed. Water soluble gases are removed from the air stream, absorbed through the mucous membrane that lines the nasal passage, and passed into the blood stream. Removal approaches 100 percent for hydrogen chloride, ozone, sulfur dioxide, and ammonia, but little nitrogen dioxide is removed and scarcely any hydrocarbons or carbon monoxide.

The nose filter can be bypassed by the mouth breathing common in people with irregularities in the bony structure of the nose, in children with enlarged adenoids, and in people of all ages when exercising vigorously or when upper respiratory infections cause congestion of the nasal passage. Oddly, the sulfur dioxide removed so efficiently in the nose at high concentrations, seems to slip through and penetrate the lung at lower concentrations[7]. The gas reaches deep into the lung in another way: It is adsorbed onto the tiny particles which are carried into the bronchioles and alveoli. The latter mechanism also carries other gases deep down.

As the warmed, moistened, filtered air moves down the trachea and into the bronchi, it carries particles small enough to escape the nasal filter, but some of these simply remain suspended in the air and are breathed out again. In passing over the mucous blanket which lines all the large air passages down to the bronchioles, many of the remaining particles may be trapped. The mucus rests on cilia like those in the nasal passage. They beat in coordinated waves about a thousand times a minute and escalate the mucous blanket with its entrapped impurities to the upper air passages where it is expectorated or swallowed. Infectious microorganisms (bacteria and viruses) or particularly irritating particles, surrounded by mucus, may be expelled by coughing. But the cough reflex, originally called into play as a defense, may itself have an irritating effect or if coughing becomes chronic, the reflex may be weakened to the point that it can no longer effectively expel mucus or particles.

Expulsion of unwanted substances can be retarded if the cilia are unable to move the mucus; sulfur dioxide, nitrogen dioxide, and cigarette smoke all slow this action. The longer irritant particles remain in the mucous blanket close to the lining cells, the greater the chance for injury and disease. When an individual inhales irritants or suffers a respiratory infection, certain cells respond by producing additional mucus. The mucus may become so thick and copious that it drowns the cilia and makes them ineffective. In people with chronic bronchitis, this becomes a permanent condition.

Particles that penetrate all the way to the alveoli—tiny air sacs at the ends of the air passages—may be engulfed by scavenger cells (phagocytes).

Enzymes in the phagocytes permit them to digest and liquefy some particles, as they do with the fragments of dead cells which are disposed of constantly. Although the body always has a reserve supply of phagocytes and produces more as needed, this response appears to be effective only up to a certain level. If the dust load to be disposed of becomes too heavy, the response is overwhelmed and some of the dust remains in the alveoli or collects at the intersections of the airways. Cigarette smoke has a toxic effect on scavenger cells. An inorganic mineral particle cannot be destroyed by a phagocyte that has taken it in; on the contrary, the phagocyte may be destroyed by the particle. With the death of one cell, the particle may be again taken in, only to kill anew. Particles may become lodged in the intercellular tissues, either to remain there or to be carried away as these tissues are drained by the lymph flow. In the latter case, the particles may become lodged in the lymph nodes (small collections of lymphocytes which appear at frequent intervals along the course of the lymphatic vessels) or may enter the blood stream when the lymph flow empties into the veins. The induction of the enzymes which make the scavenger cells work may be inhibited by carbon monoxide, thus lowering the effectiveness of the lung's defenses; certain biochemical processes within cells may be depressed by the inhalation of lead[8].

Still other lines of defense against foreign toxins are the body's ability to produce antibodies which can neutralize toxins in the blood stream and its ability to produce interferon which aids in resistance to viruses. A recent study has found that carbon particles, sulfur dioxide, and nitrous oxides interfere with the formation of antibodies when infectious bacteria are administered to test animals[9]. Finally, waste materials in the blood may be metabolized in the digestive organs and eliminated through the kidneys.

The wider the air passages, the more easily air can flow in and out and the more easily we can rid ourselves of mucus. Some air pollutants, most notably sulfur dioxide, nitrogen dioxide, and ozone, may cause spasms of the muscles of the bronchial airway and edema or swelling of the lining membrane. The net result is a narrowing of the airway and interference with both the passage of air and the removal of mucus. Just how this works is not clear, but it does not appear to require direct contact of the gases with these air passages; that is, the nasal defenses may be effective in removing the gases from the inspired air, thus ensuring that they do not come into *direct* contact with the lower respiratory system, but the gases are nevertheless able to act on that system *indirectly* by way of the bloodstream.

Other Parts of the Body. This is one demonstration that the transfer of gases from the respiratory tract into the blood stream does not eliminate their capacity for harming the organism. The thin walls of the capillary blood vessels must be easily permeable to oxygen from the lung and to carbon dioxide from the blood in order to carry out respiration. This makes the walls easily permeable to carbon monoxide, also. Once in the blood, carbon monoxide combines with the hemoglobin of the red blood cells whose function it is to transport oxygen to the tissues; it combines with hemoglobin so much more readily than oxygen that the oxygen is displaced, thus interfering with the need of all body cells for a steady supply of oxygen. Fortunately, carbon monoxide can also pass through the capillary walls from the blood to the lung where it is expired. If exposure to carbon monoxide stops, the blood will be cleared slowly—about half the carbon monoxide already in the blood disappears in three to four hours. While carbon monoxide remains bound to the hemoglobin (as carboxyhemoglobin), it is carried to the tissues in place of oxygen and affects the myoglobin (like hemoglobin, an iron-containing pigment) which stores oxygen in the muscle and the cytochromes within the cells which are the final stage in the oxygen pathway.

The oxygen needs of the body of course vary tremendously, depending upon whether it is at rest, moving about, or exercising vigorously, and the whole respiratory and cardiovascular system responds to these changing needs, most obviously by deeper breathing and a more rapid beating of the heart to increase the blood flow when oxygen needs are greatest. Interferences with the delivery of the needed oxygen—whether by narrowing the air passages, damage or obstruction in the lung, or replacement of oxygen by carbon monoxide—elicit one or more of the adaptive responses for maintaining the oxygen flow to the tissues: More oxygen will be withdrawn from the inspired air or from the hemoglobin or the myoglobin; the lungs will pump in more air; the arterial blood flow will increase; and the heart-beat rate will increase.

The fact that people can move from a low to a high altitude without harm indicates the body's ability to adapt to variations in oxygen intake. However, there are some important differences between an adjustment to a new, constant oxygen content in inhaled air and the requirements of coping with reduced oxygen content caused by internal stresses such as carbon monoxide in the blood, excessive mucus in the lungs, or physical injury to some part of the pulmonary system. Healthy nonsmokers, after inhaling carbon monoxide, showed adaptive responses in the circulatory system, but slight impairment of some aspects of lung function[10]. (The concentration of carbon monoxide inhaled was very high, but the exposure brief. The resulting carboxyhemoglobin in the blood was 3.95 percent, similar to that of people smoking about a half pack of cigarettes a day.)

Particles, too, may harm the body as long as they remain in it, even though they have been cleared from the pulmonary system. The mucus is continually swallowed and carried with its load of impurities into the gastrointestinal tract. This is necessary for the elimination of the particles, but along the way it opens the door to damage to the digestive organs.

Particles are usually thought of as solid, but a substance which first appears in the air in this form may interact with other atmospheric constituents to become part of a compound in the form of gas or vapor or, once inside the body, may be dissolved in the mucus. There are other mechanisms for carrying material across membrane barriers, but they are not completely understood. However, it is clear that there are possibilities for both particles

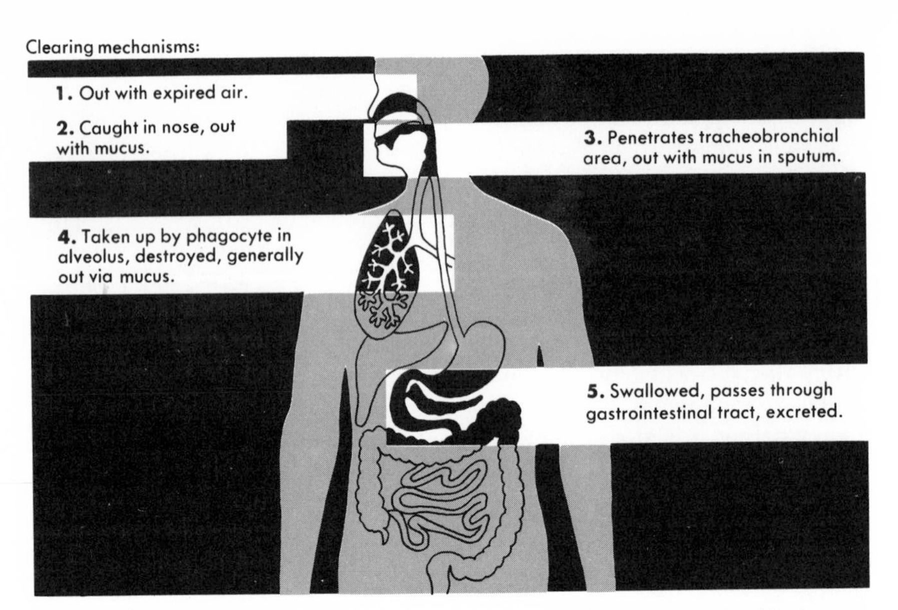

The body's defenses against invasion by foreign particles are considerable, but can be overwhelmed by excessive amounts of particulate pollution, by breathing particles in combination with gases, or by particles that are infectious, irritating, or toxic.

and gas to pass into the blood directly as well as indirectly via the lymph flow. Components of the blood itself, the central nervous system, and the bones may thus be exposed to damage from lead, cadmium, radioisotopes, or other harmful substances.

Cadmium, for example, can be carried to the kidneys where it concentrates and can produce certain destructive kidney changes. It is also bound by blood vessels and alters their reactivity. Although scientific opinion is divided, there is considerable support for the hypothesis that cadmium causes high blood pressure as well as kidney damage and may shorten life[11].

Epidemiologic evidence that air pollution lowers resistance to respiratory infection led to laboratory investigations of how this might be brought about. It was learned that nitrogen dioxide—one of the gases low in solubility and therefore not removed in the nose, but allowed to penetrate into the lower respiratory tract—lowers resistance to bacterial and viral infections. (*Klebsiella pneumoniae* and influenza virus were the infective agents used in a variety of experimental animals[12].) Ozone has been found to impair the ability of phagocytes to trap streptococci[13]. Inhalation of lead oxides can also interfere with pulmonary phagocytes[14], and so can loading the respiratory tract with dust[15].

Certain kinds of particles, if not removed from the lung, may become surrounded by dense, fibrous tissue. Nodules of such tissue may be found throughout the lungs if exposures to the irritant dust are high; these lead to impairment of respiratory function, sometimes to other diseases of the lung,

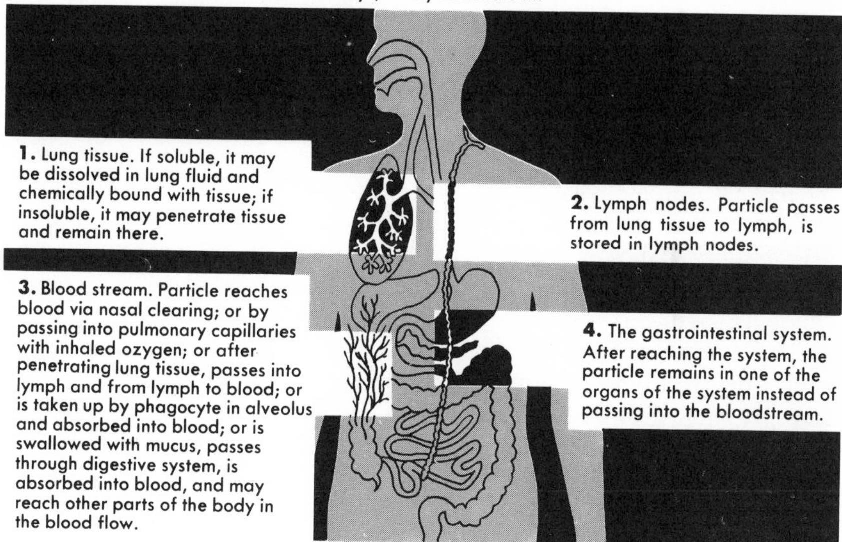

and ultimately to death. Silicosis, caused by the inhalation of dust containing a high proportion of silica, is the best known of the chronic diseases of the lung arising from dust exposure, but there are a number of others collectively referred to as the pneumoconioses. These have always been considered occupational diseases: asbestosis of asbestos workers; coal-miners' pneumoconiosis or black lung; hemp and cotton-mill workers' byssinosis; silicosis in miners—gold miners, coal miners, and others—and in granite workers, sand-blasters, and workers in the processing of metals with a high silica content, such as foundry workers; berylliosis in workers in plants using beryllium, such as weapons factories producing rocket propellants; farmers exposed to grain dusts. Only recently has it been recognized that some fibrous dusts are present in ambient air and in the lungs of those who breathe it. Some of the same dusts may cause cancer as well. Asbestos can induce mesothelioma, a cancer of the pleural cavity.

Asbestos in the air of New York City, Philadelphia, and suburban Ridgewood, New Jersey, was found to vary from one-hundredth to one-tenth of a microgram per cubic meter. Urban air typically contains particulates in amounts from 75 to more than 200 micrograms per cubic meter of air, so even one-tenth of a microgram of asbestos in a cubic meter of air appears to be very little. However, asbestos easily fragments into fibrils so small that a thousandth of a microgram of this material could represent a million fibrils. A common practice in the building of high-rise office buildings is to spray fireproofing material containing from 10 to 30 percent asbestos onto the girders and other structural components. Not only are workers engaged in the spraying or otherwise occupied on the same building exposed to the spray, but intensive fallout in the vicinity can often be seen. Ambient concentrations as high as 0.18 micrograms have been measured near such a source. The national emission standards for asbestos proposed on December 7, 1971, by the Environmental Protection Agency prohibit asbestos spraying in the future, and some cities already restrict it.

Asbestos bodies have been found at autopsy in the lungs of city dwellers who had no occupational exposure in South Africa, Canada, Italy, England, Scotland, Ireland, and Germany, and in Pittsburgh, Miami, and New York in this country. Asbestos bodies were found in a small sample of lung tissue of 1449 out of 3000 consecutive autopsies in New York[16].

Asbestos is only one of the fibrous materials that may be present in community air; whether these materials are present in sufficient quantity to represent a danger to the most susceptible segments of the general population is not yet known. The extent of the problem that awaits investigation was summarized by Lewis J. Cralley at the International Conference on Pneumoconiosis:

> There are well over 100 different natural minerals with some degree of fibrous structure. They may be present and dominant in a number of ores of commercial value or may be present generally in areas of the earth's crust, in fossil fuels, and in other commercial minerals. They may be disseminated into the air through the action of natural forces such as dust storms [or] in the mining, processing, and subsequent use of the minerals as fillers and carriers of reactive chemicals. Their occurrence may relate to personal habits, such as the use of cosmetic talcum powder which contains a high percentage of respirable fibers[17].

Cancer may also arise from inhalation of certain gaseous or particulate pollutants, alone or in combination. Most attention has been focused on lung cancer, which has risen spectacularly in the twentieth century and is more prevalent in cities than in rural areas. Although some of the rise may be apparent rather than real, reflecting improved diagnosis, much of it is a real increase probably due in part to increased smoking, in part to occupational exposure, and in part to community air pollution[18].

The federal criteria document on particulate matter noted that "The possibility of relatively high concentrations of an atmospheric pollutant appearing in organs remote from the lungs must not be overlooked."[19] And, in fact, an association between air pollution and stomach cancer has been found in England[20], in Nashville[21], and in Buffalo[22]. Selikoff has found cancer at other sites in addition to the lung in asbestos workers[23].

The cell damage leading to cancer has been under intensive study for many years, but exactly how it occurs is not yet fully understood. Once the damage is done—for example, if a chromosome is broken—it may be repaired, but the repair mechanism also is not fully understood. It has been found that some pollutants will cause cancer in experimental animals only in combination with other pollutants, or more readily in combination with others, and it may be that in such cases one pollutant interferes with repair.

Some pollutants cause cancer only in combination with other pollutants or more readily in combination with others. The hazard of occupational exposure to air pollutants has been frequently underestimated and is particularly dangerous in combination with cigarette smoking. Uranium miners who smoke are in greater danger of developing lung cancer than either nonsmoking uranium miners or smokers with no occupational exposure to radiation.

When animals have been infected with influenza virus and exposed to artificial smog (ozonized gasoline), lung cancer has been produced [24]. In any case, cancer induced by exposure to two pollutants or to an infectious agent and a pollutant further underscores the error in trying to establish a one-to-one correspondence between a given concentration of a single pollutant and a certain health effect.

The urban air pollutants most suspect as contributors to cancer induction are the polycyclic hydrocarbons such as benzpyrene, part of the organic fraction of particulate matter. They are present in cigarette smoke as well. Tumors have been induced in laboratory animals by exposure through injection or implantation of benzpyrene and by inhalation of road dust and soot. In the latter case, the combination of polycyclic hydrocarbons and inert dust, both present in the road dust, appeared to be responsible for the tumors [25]. Inhalation of high doses of sulfur dioxide by itself has produced tumors in laboratory animals[26], while lung cancer like that in humans has been induced in animals for the first time by inhalation of benzpyrene in combination with sulfur dioxide[27]. Other pollutants which sometimes appear in urban air and have been demonstrated to be carcinogenic by themselves are nickel carbonyl and beryllium.

In both occupational medicine and in the laboratory, the importance of the time factor in the operation of the body's defenses has been observed. Clearing the lung, exhalation of the carbon monoxide in the blood, excretion of the pollution burden in the gastrointestinal tract, and cellular repair all require time, so that interruption of exposure is very important in permitting these defenses to work. The continuous nature of air pollution allows less scope for detoxification and excretion.

Genetic Changes and Adaptations. Cell damage may affect the genetic material, causing mutations which are passed on to future generations. If an external agent injures one of the chromosomes in a cell anywhere in the body, the replication of that cell may be affected and—unless the injury is repaired—this may be the first step toward the abnormal cell growth of cancer. If the injured chromosome is in the egg or sperm, it may result in a birth defect in the child or in one of its descendents.

Chromosome breakage can be observed in microorganisms, experimental animals, or human tissue, and may serve as a warning (though not a proof) of the possibility of trouble in future generations. However, if only one of the many genes on the chromosome is damaged, resulting in what is called a point mutation, the chromosome may not be visibly affected and succeeding generations must be observed before the effect is seen. More study of the possible mutagenic effect of air pollutants is needed, but is extremely difficult, especially in the case of point mutations[28]. In the laboratory, however, unsettling information is developing. Sulfur dioxide has been found to be transformed in body fluids to sodium acid sulfite; the sulfite brings about further biochemical changes in the chromosomes. Mutations have been produced in bacteria by sulfur dioxide pollution[29]. Radiation is known to cause mutations, and the discovery that ozone is *radiomimetic*—that is, that it "mimics" the mode of action and biological effects of radiation, winding up in premature aging and chromosomal injury—makes ozone suspect as both a carcinogenic and mutagenic agent[30].

There is now general agreement that the intense exposure to air pollution that resulted in 4000 deaths in London in 1952 must not be allowed to recur (although we are not yet moving fast enough to prevent it). What to do about less intense exposures is more controversial. It has occasionally been suggested that the human body's remarkable adaptability to a variety of environments—the way it can become acclimated to changes in altitude and temperature, for example—might extend to some level of air pollution. A symposium on health hazards in man's environment which considered various aspects of adaptation, however, found "little evidence of physiologic adaptation to ionizing radiation or to air pollution."[31]

Scientists have learned that in the rare cases where laboratory exposure appeared to convey some protection against succeeding exposures, this protection was conferred at a biologic price. After exposure to ozone, mice developed a tolerance to it which protected them against the acute effects of further doses, doses which injured or killed control mice not previously exposed. The mice which had developed the tolerance were also protected against other pulmonary irritants such as nitrogen dioxide, hydrogen sulfide, and phosgene. However, as Herbert Stokinger and D. L. Coffin[32] point out after summarizing the experimental evidence, pretreatment with ozone does *not* provide protection against all toxic effects of succeeding doses of ozone and other irritant gases; protection is afforded only against edema (flooding of the lung with watery fluid), an acute effect of these irritants. They add that other experimental evidence "leads one to suspect that tolerance may actually initiate some chronic toxicologic effects."[32] In other words, the defense against acute effects produces its own stress on the organism.

The difference between genuine physiologic adaptation and what should more properly be called adjustment has been pointed out by René Dubos[33]. Once we have reduced exposure to environmental contaminants to levels which eliminate the kind of toxic effects that are immediately disabling and otherwise obvious, human beings, says Dubos, may become accustomed to pollution levels that they do not regard as a serious nuisance and that do not interrupt social and economic life. But it is probable that continued exposure to low levels of toxic agents will eventually result in a great variety of delayed pathologic manifestations. The point of importance here is that the worst pathological effects of environmental pollutants will not be detected at the time of exposure; indeed they may not become evident until several decades later. One of the reasons we have been slow to recognize air pollution's threat to health is that in chronic bronchitis, the pneumoconioses, and cancer there is a long period between initial exposure and full-blown disease. Society may become *adjusted* to pollution levels that do not create an immediate nuisance. But the human body will not become *adapted* to these levels of pollution in the sense of being able to inhale them without harm. The *apparent* adaptation will eventually cause much pathological damage in the adult population and create large medical and social burdens[33].

In the case of mutagenesis the time lag is still greater. If the genetic material is injured, producing a mutation which is dominant (appearing when the causative gene is present singly), the mutation may be observed in the next generation. But if the mutant characteristic is recessive, it will appear only in the offspring of parents who each carry the mutant gene, and may not appear until several generations later when an increase in the mutation rate might be noted but the cause lost in the passage of time. The genetic testing of microorganisms, which can go through many generations in a day, can be helpful. The relationship of such tests to man has been well expressed by James F. Crow:

> The most sensitive tests of mutagenicity employ microorganisms; hence it is no surprise that there are a number of chemicals that are clearly mutagenic in these organisms, but whose effect on higher organisms is unknown. Nevertheless, identity of the genetic material in all organisms implies that a chemical that is mutagenic to one species is likely to be in others and must be viewed with suspicion[34].

All the foregoing mechanisms for breaching the body's defenses either have been observed in people or have been observed in laboratory animals with reason to think that the same *mechanism* would operate in people, although not necessarily at the same level of exposure to pollution. In no case is it possible to specify a pollutant exposure below which one could be certain that any given defense would operate successfully. One reason for this is that the threshold of safety is not the same for everyone, nor the same at all times for any given individual.

From Birth to Death in Polluted Air

The federal criteria document for carbon monoxide[35] points out one way in which the unborn child may be sensitive to pollution. Carboxyhemoglobin levels in maternal blood, reflected in umbilical cord blood, are higher in mothers who smoke than in nonsmoking mothers. Exposure to carbon monoxide in community air, as well as in cigarette smoke, could raise the carboxyhemoglobin in the maternal, and thus in the cord, blood. Whether the higher carboxyhemoglobin in itself results in any other effect on the fetus is at this time uncertain, the criteria document continues, but unborn infants who may have exceptional requirements for oxygenation during childbirth "may well be among the population at high risk."[35]

There are physiologic reasons for considering that a child is likely to be particularly susceptible to air pollution during the first weeks and months of life. In spite of resistance to infection conferred by antibodies from the mother, newborn infants are especially susceptible to bacterial infection. The synergistic effect of respiratory infection and air pollution has been mentioned. Infant lungs are of course very small: Their aerating surface is in fact smaller relative to body weight than that of an older child or an adult because the tiny air sacs at the end of the airways are just beginning to develop at five weeks of age. An eight-year-old child, on the other hand, has a full complement of 300 million fully developed alveoli. The same concentration of a pollutant might therefore represent a greater hazard to the infant with his more limited aerating surface. The infant lung is also immature in a number of other ways: The structure is simpler, the mucus glands have not yet developed in the small air passages, and the supporting elastic fibers and muscular tissues are present only in the larger bronchi. These factors make clearance of mucus and foreign matter more difficult and permit infection, once established, to spread more rapidly than in a fully developed lung. There is some epidemiologic evidence that infants may be at special risk. For example, in the worst week of the London disaster of 1952, while mortality for all ages increased, mortality for newborn infants doubled and more than doubled for infants one to twelve months old[36]. In the laboratory, newborn mice were exposed to particulate pollutants from the air of American cities and, during adult life, many of these animals exhibited various forms of cancer[37].

In older children, respiratory illnesses range from the common cold to the severe viral and bacterial infections. There have been a number of studies of the effects of air pollution on children that have shown that impaired lung function is more frequent and more marked, and there is more respiratory illness in children living in polluted areas than in those breathing cleaner air[38]. Increased respiratory illness in children with greater exposure to air pollution has not been found in Los Angeles where photochemical smog is the main problem, although it has been looked for there, but it has been found where nitrogen dioxide is the dominant pollutant[39] and in the sulfur and particulate pollution of British cities[40].

A British study followed children born in 1946 until they were 15 years old[41] Not only did this study find frequency and severity of lower respiratory infections increasing with the amount of air pollution exposure at all ages, but by age 15 persistence of chest noise suggestive of congestion was ten times as prevalent in the high pollution area as in the low.

D. D. Reid, one of Britain's—and the world's—leaders in air pollution epidemiology, pointed out some implications of respiratory mortality in "The Beginnings of Bronchitis."([40]) A progression from respiratory disease in childhood to chronic bronchitis in adulthood is certainly not inevitable. On the other hand, says Reid, in generations which suffered relatively high respiratory death rates (for males) in childhood, high respiratory death rates also appear when survivors reach middle or later life.

> This implies that males exposed during infancy and childhood to epidemic conditions causing heavy respiratory mortality continue to be especially predisposed to mortal lung disease throughout life[42].

When British and Scandinavian respiratory mortality is compared, a similar excess of deaths is found in Britain in both childhood and middle age.

> This consistency implies that some etiological factor or factors in the national environment affects respiratory disease in both these age groups[43].

Within Britain, there is an excess of respiratory mortality in infancy and at ages 45–64 in urban as compared to rural areas.

Among adults, heavy smokers are much more prone to develop chronic bronchitis and progressive, irreversible emphysema than are nonsmokers, and men are more susceptible than women[44]. But both smokers and nonsmokers in highly polluted areas are more prone to these diseases than their counterparts where the air is relatively clean. Bronchitis is caused by irritation of the bronchial wall lining and is marked by coughing and excessive mucus production. In emphysema, connective tissue in the lung is deranged or destroyed. There is a loss of lung elasticity and obstruction to air flow and a constant shortness of breath. The death rate from emphysema has doubled every five years for the last 20 years. As with lung cancer, improved diagnosis may account for some of this increase, but most is considered real by specialists in respiratory disease.

In studies of London postmen, those delivering mail in more polluted areas were found to have much higher rates of acute illness, chronic disability, early retirement, and death from chronic bronchitis, emphysema, and related heart disease than those in cleaner suburban areas[45]. Some of these men reported that respiratory illness had caused frequent and prolonged absence from work even when they were in their early twenties.

Once an adult has chronic bronchitis, his symptoms are likely to be intensified if pollution rises[46]. People with respiratory (and circulatory) diseases are adversely affected by high pollution in the photochemical-smog-laden atmosphere of Los Angeles, as well as in cities where pollution from industrial processes and coal burning predominate[47].

From London we are now beginning to receive indications of *reductions* in respiratory disease with *improvements* in air quality. The improvements have been primarily in the reduction of particulates. Death rates from bronchitis and, to a lesser extent, from lung cancer are falling, and other indices of bronchitis also indicate improvement[48].

One of the greatest difficulties in seeking the relationship between air pollution and death is in finding a way to hold all other factors constant so that the results of air pollution alone can be measured. This has been most satisfactorily done in analyses of episodes of very high air pollution like those in London in 1952 and in New York in 1962. The number of deaths in those cities that would be expected in the same number of days, at the same time of year, with temperatures about the same, were calculated from experience in previous years. The expected deaths were compared with actual deaths during the pollution episode and the "excess deaths" attributed to air pollution. But the "normal" periods that were compared with the episodes were not periods of *no* pollution, but merely periods of what had become "normal pollution" in those cities. The studies left unanswered the question of whether people were dying from the effects of air pollution at these normal levels.

Bertram Carnow undertook to study deaths in Chicago during one month, recording deaths for periods of high, medium, and low pollution (rated according to sulfur dioxide levels), comparing deaths within each square-mile area to determine whether changing pollution levels affected mortality rates, then looking at the city as a whole. Citywide sulfur dioxide levels in the "low" third of the month averaged 0.08 parts per million, considerably above the present national standard of 0.03. The medium third of the month averaged 0.17 parts per million and the high third, 0.24. In some parts of the city, sulfur dioxide levels rose to more than 0.34. Other pollutants, of course, were also present. No significant rise in mortality was

London on December 6, 1952, during the period of heavy smog which was blamed for 4000 deaths.

Although London has since made great progress in cleaning up urban pollution, it has been growing elsewhere—in Birmingham, Alabama, for example, where there were episodes of extremely high particulate concentrations in April and again in November of 1971. This photograph is from the November episode, during which 23 large industries were required to shut down briefly.

noted for women at higher pollution levels, but deaths among men due to cardiovascular and respiratory disease showed a striking rise[49].

Males 55 and over appeared to be the high-risk group. In this group, there was no change in average deaths per day between the low and medium pollution periods, but there was a significant rise between the medium and high pollution periods. Most interesting was the pattern that was revealed when high and low socioeconomic areas were observed separately (still with reference to males 55 and over). The rise in average deaths per day was not uniformly associated with any given pollution level. Among the well-to-do, it showed up when pollution rose to 0.24; among the poor, not until it rose to 0.38. When areas of predominately white residence and those of predominately black residence were considered separately, the same pattern was revealed; it required a greater increase in pollution to produce an increase in mortality among black males 55 and over than among white males of the same age group. As Carnow pointed out:

> There are a number of possible explanations for these patterns. The first is that low socioeconomic or nonwhite individuals have increased resistance to disease and therefore show significant mortality only at very high sulfur dioxide levels. The higher death rate in these groups, the poor nutrition, the lack of available medical care, poor housing, and other factors suggest that this is not the case. The second possibility is that adaptation occurs to a higher degree at the low socioeconomic level and in nonwhite individuals. The increased death with the increase in pollutants suggests, however, that this is not so[49].

Carnow then proposed an alternative explanation—namely, that the most susceptible individuals in all population groups are killed by a combination of illness and environmental factors including *the usual levels of pollution* in the areas where they live, but, since this happens all the time, their deaths appear as part of the normal death rate. This points up the difficulty of ever really pinning down the number of deaths for which air pollution is responsible. An individual may reach the point at which he is at risk of death from pollution because of many things, including the effects of air pollution on his health in previous years.

Populations at Special Risk

Individuals differ in many ways, not the least of which is in the condition of their defenses against air pollution. Some have inherent deficiencies which make them particularly susceptible; others have preexisting disease which has lowered their resistance; and still others have a background of special exposure which makes them, at any given time, more liable to harm from additional exposure.

Asthma. Among those with respiratory defects are individuals whose bodies produce allergic antibodies which, instead of playing a detoxifying role, combine with an allergen taken into the body. Such individuals may be allergic to pollen or to some other airborne pollutant. The antibody-allergen combination results in the release of mediating substances. Of these, histamine is the best known to laymen due to the wide use of antihistamines in the treatment of allergies and the common cold. [There are also serotinin, SRS (slow-reacting substance), and other mediators.] These mediators in turn act on blood vessels, mucus glands, and smooth muscle in the lung to cause smooth muscle contraction, swelling of the lining of the bronchial tree, and production of thick, tenacious mucus. The individual wheezes, coughs, and grows short of breath—he experiences an attack of asthma. But exposure to a pollutant may result in an attack of asthma in an asthmatic individual although neither the allergen nor the antibody can be identified. In this instance the pollutant acts not as an allergen, but as an irritant.

Air pollution has been directly related to increased asthma attacks in a number of studies:

> There were increased emergency room visits for persons with asthma on the third day of the 1966 air pollution episode in New York[50].
>
> A direct correlation was found in Nashville between sulfur pollution and the number of asthma attacks in adults. Here again there was a lag, with the asthma attack rate being higher a day after the exposure to increased air pollution[51].
>
> During the 1948 disaster in Donora, Pennsylvania, 90 percent of the asthmatics were affected, while 40 percent of the total population was affected[52].
>
> A study of Philadelphia school children found a threefold greater incidence of asthma during days of severe air pollution. When this was combined with an increase in barometric pressure, the number of asthma attacks increased ninefold[53].
>
> Hospital admissions of children in Buffalo and environs with chronic allergic disease—asthma in the period 1956–1961 and eczema during 1951–1961—were studied in relation to average air pollution as measured by particulates in the area where the children resided as well as in relation to their socioeconomic class. The investigators report that "Incidence rates rose steadily within each social class with the increase in air pollution. Conversely, there was no apparent relationship between social class and the incidence of hospitalized cases."[54] It might be added, however, that where a child lived, and therefore how much pollution he was exposed to, was a function of his socioeconomic class. Seventy-eight percent of the child population under 15 years of age in the lowest class in Buffalo lives in areas with particulate averages of 100 micrograms per cubic meter of air or more, while 90 percent of the child population in the highest class lives in an area with particulate averages of less than 100.

Chronic allergic disease is very common in childhood. It takes the form of hay fever, eczema, or asthma, but an allergic child may suffer from all three forms. Some children outgrow the disease; in others it persists throughout life, sometimes changing its form. Adults who did not suffer from it in childhood may begin to experience asthma in their forties or later. Although air pollution has been found to increase the incidence of asthma *attacks,* there has not yet been a study of its effect on the onset, the incidence, or the persistence of the disease. However, it is known that the incidence of asthma has increased in urban areas[55] and that persistence of asthma over a long period may lead to progressive irreversible emphysema. The wider implications of air pollution for this susceptible part of the population therefore need to be explored.

Antitrypsin Deficiency. In the complex destructive processes by which the body breaks down food and worn out cellular material, enzymes are the catalysts. To keep these processes from destroying living tissue that is still necessary to the body, there are regulators of enzymatic activity. One such regulator is alpha-1 globulin antitrypsin, an inhibitor of the enzyme trypsin. Some people have a hereditary deficiency of alpha-1 antitrypsin and, in recent years, it has been learned that people with such a deficiency are particularly susceptible to emphysema[56]. It is not clear how the antienzyme deficiency leads to emphysema—whether directly, through increasing the rate of degradation and replacement of the connective tissue or through altering the connective tissue proteins, or indirectly, by increasing susceptibility to infection, or by some other mechanism. Neither is it known why the lung should be the organ affected, but it is quite likely that people with this deficiency are more susceptible to air pollution than others as they have been found to be particularly susceptible to cigarette smoke[57]. It has recently been suggested that a study of the effects of air pollution on this sensitive population group might tell us more about the effects of low levels of pollutants[58].

Heart and Circulatory Problems. Since the heart and the lungs are an interconnected system, agents which injure the lungs could be expected to put a strain on the heart that is secondary to the lung damage, if not directly injurious to the heart. As we have seen, cardiovascular disease has accounted for many deaths in air pollution disasters. In the Nashville study, there were found to be more deaths from heart disease in highly polluted areas than in cleaner areas[59].

One heart problem—*cor pulmonale*—is in itself an outcome of respiratory disease. When airways have been constricted or blocked by excessive, thick mucus or fibrous nodules or when the lung is rendered abnormal by emphysema or infection, pulmonary blood pressure rises. The right side of the heart works harder to continue the normal flow of oxygen to the cells, and may eventually fail under the work load.

People with circulatory problems that are not necessarily an outcome

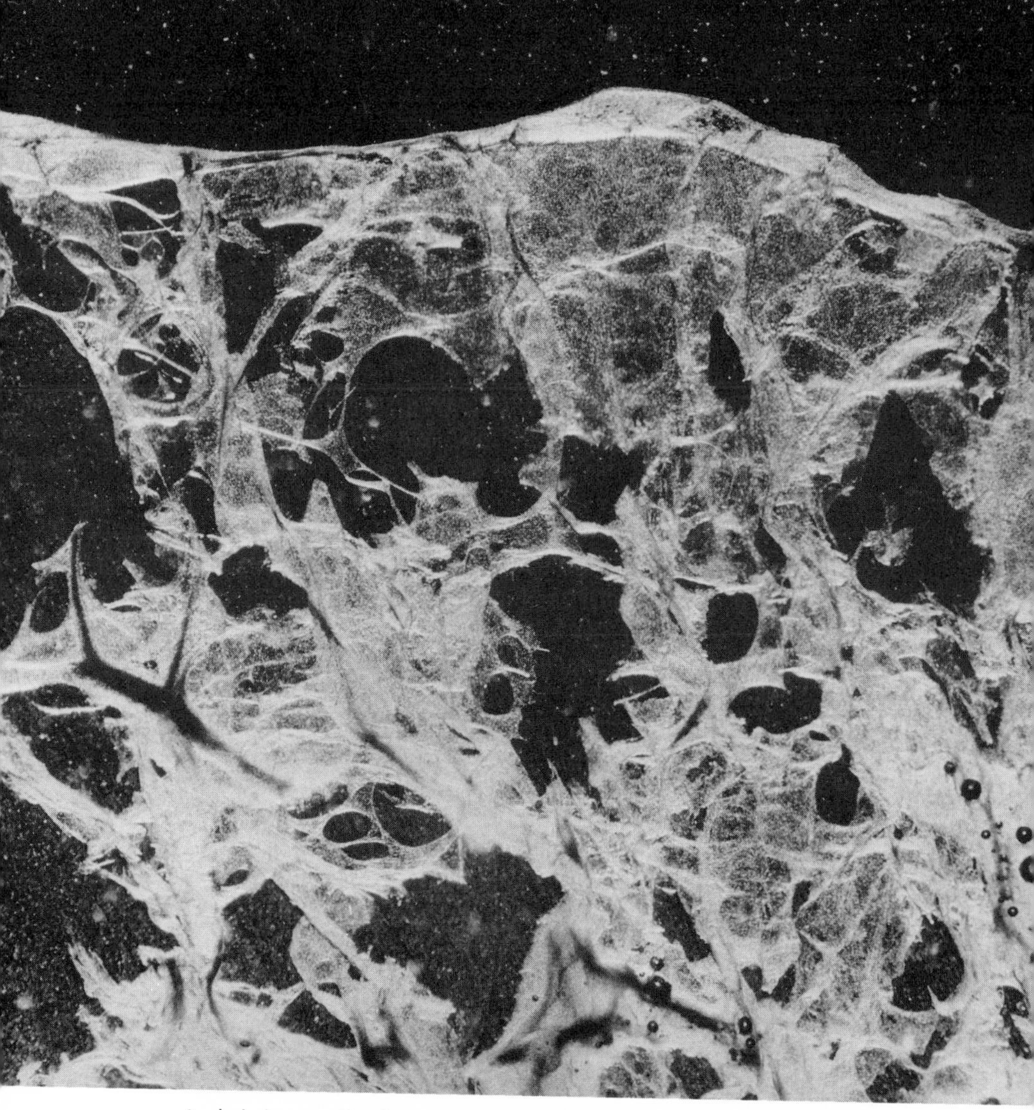

A whole lung section from a patient with alpha-1 antitrypsin deficiency who died of emphysema. People who suffer from this enzyme deficiency are particularly susceptible to emphysema. Smoking presents a special hazard in their cases, and they may be particularly susceptible to air pollution as well, although this has not been studied.

of pulmonary disease, when exposed to high levels of carbon monoxide and a consequent reduction in the oxygen-carrying capacity of the blood, could also be expected to suffer. An association has been found between high levels of carbon monoxide in Los Angeles air and increased mortality[60] and, in particular, between high carbon monoxide and myocardial infarction[61]—an acute coronary condition in which a blood vessel is stopped up by an obstruction, the blood supply to part of the muscular substance of the heart is shut off, and the deprived tissue dies.

Anemia. People who suffer from anemia, whatever its cause, have an abnormally low supply of oxygen-carrying hemoglobin and are therefore likely to be especially sensitive to an increase in carbon monoxide which interferes with hemoglobin.

Lead has many and varied effects on health, but the most common is anemia. There is both interference with red blood cell formation and injury to mature red blood cells. If the exposure ends, the production of red cells in the bone marrow offsets the anemia; if the exposure continues, the defensive response cannot fully compensate for the damage.

People who are already anemic are handicapped in dealing with an invasion of lead because they are unable to step up their production of red blood cells. Among such people are groups of substantial size who suffer from hereditary anemias. The genetic trait for sickle-cell anemia afflicts an estimated ten percent of black people in this country. Those who inherit the trait from both parents are anemic, usually develop serious health problems, are susceptible to infections, and seldom live past age 40. They may be at additional risk from the carbon monoxide in cigarette smoke and in ambient air and from exposure to lead.

Those who inherit the trait from only one parent may be unaware of it and apparently healthy, unless a shortage of oxygen precipitates a problem. During World War II, soldiers with the sickle trait were unable to withstand the shortage of oxygen in unpressurized aircraft at high altitudes. Theoretically, pollutants affecting the oxygen supply to the blood might be considered a special risk to people with the sickle-cell trait, although probably not without intense occupational exposure or a combination of pollution and other health-threatening circumstances.

Thalassemia, another hereditary anemia, is especially prevalent in people of Mediterranean origin. There are also one to two million people in the United States with a hereditary deficiency of an enzyme important in the maintenance of red blood cells (glucose-6-phosphate dehydrogenase)[62]. This enzyme also has an important function in scavenger cell activity, and people with this enzymatic deficiency may therefore be a high-risk group for other pollutants as well as lead and carbon monoxide.

Personal and Occupational Exposure

Smoking has appropriately been called personal air pollution. Laboratory and epidemiologic studies have implicated it in some of the same kinds of physiologic damage and in increased illness and death from the same diseases as community air pollution. But the individual, knowing the risk involved in smoking, can choose to smoke or not to smoke, thus striking his own balance of personal pleasure against that risk; he has no choice when it comes to air pollution.

When a population is divided into two areas of residence, one characterized by high and the other by low air pollution levels, and within each area into smokers and nonsmokers, both smoking and pollution affect the smokers. The figure shown here does not indicate the possible contribution

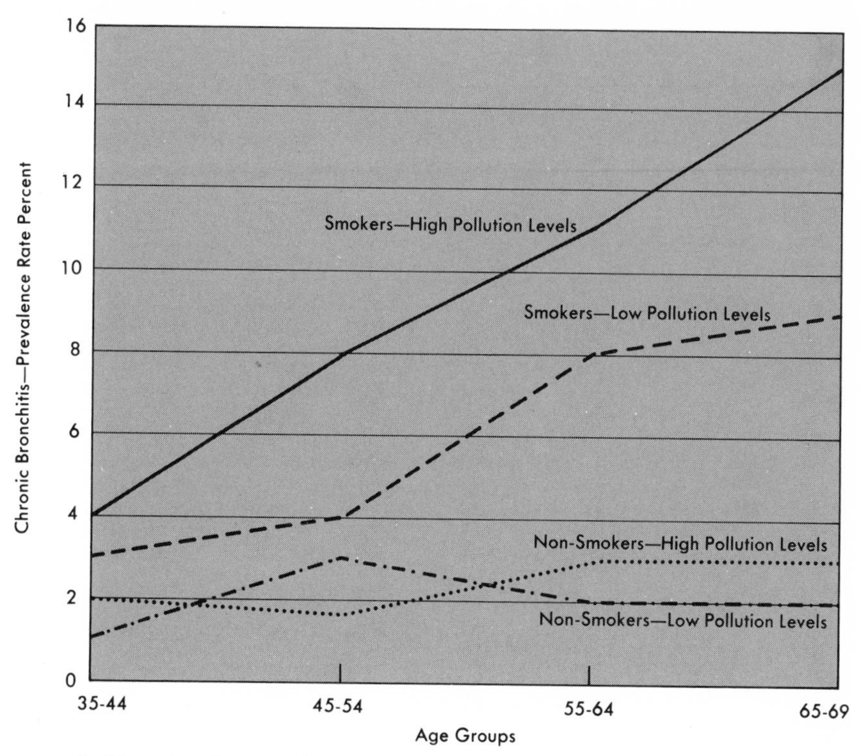

Ambient air pollution and smoking may cause respiratory disease separately or may reinforce each other. This was shown in a British study of men and women aged 35–69 which found the prevalence of symptoms of chronic bronchitis rising with increasing levels of air pollution among both smokers and nonsmokers. The widening gap between smokers and nonsmokers as pollution increased suggests that the combined effect of pollution and smoking is greater than the sum of their separate effects.

of a third type of pollution exposure—the occupational. Occupational and personal air pollution together have proved to be a lethal combination for workers in more than one industry.

Among 87 asbestos insulation workers with no history of cigarette smoking who were followed by Selikoff and coworkers from 1963 to 1968[63], only one died of lung cancer; among 283 cigarette-smoking workers in the same industry, 27 died of lung cancer and three more had cancer, but were still living in 1968. Selikoff estimates that an asbestos worker who smokes cigarettes has 90 times more chance of dying of lung cancer than a man who neither works with asbestos nor smokes cigarettes[63].

The radioactive gas to which uranium miners are exposed is known to be carcinogenic. Uranium dioxide dust to which uranium mine and mill workers are exposed may also be a factor in their total radiation exposure[64]. Tobacco smoke is also carcinogenic. The combination appears to be worse than either radiation or smoke alone[65]. This has a significance beyond the one occupational group for, together with laboratory experiments using radiation and chemical carcinogens, it suggests the importance of the interaction between the two. This is a problem in need of further investigation because of the increases in both radioactive and chemical contamination in the past quarter century[66].

Occupational exposure has almost certainly been underestimated as a factor in respiratory disease. Workers in some industries are exposed to high concentrations of the polycyclic organic matter that is suspect as a carcinogen in both urban air and cigarette smoke, but there have been few investigations of the incidence of lung cancer in such occupations. One study of the British Gas Works Industry found that heavily exposed workers

Adverse health effects often go unnoticed for a number of years in community exposures, as well as in occupational exposures. The sharp climb in the incidence of asbestosis among workers who cover pipes with asbestos begins after ten years of exposure. As can be seen here, almost 40 percent of the workers exposed more than 20 years had symptoms of asbestosis.

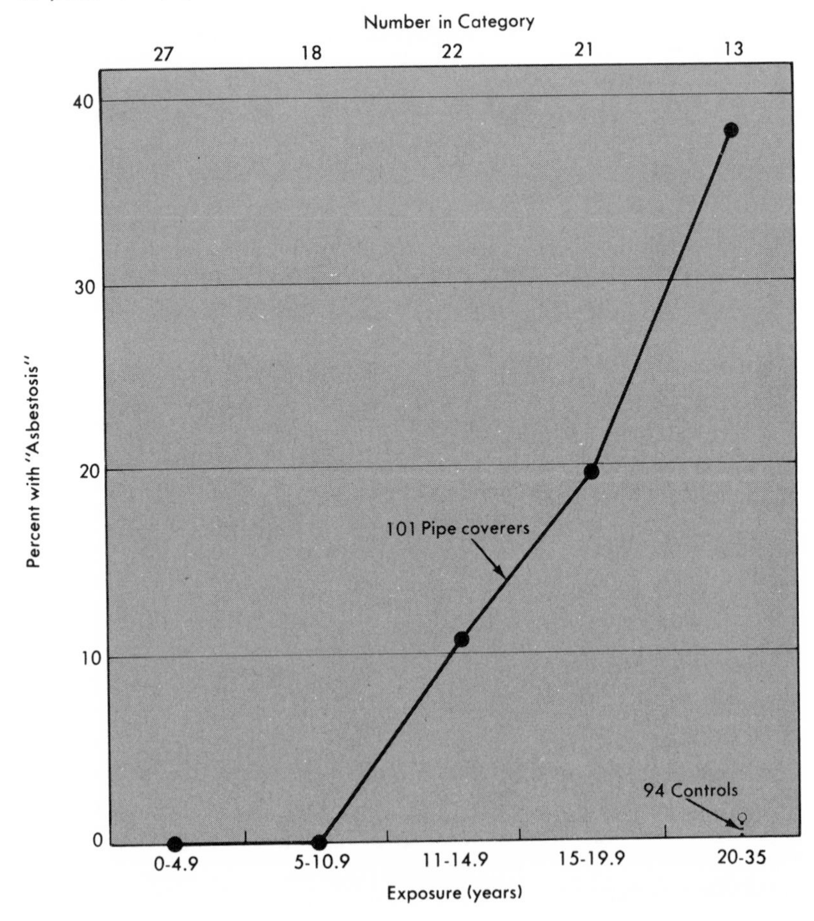

had a 69 percent higher lung cancer incidence and a 126 percent higher death rate from bronchitis than workers with no exposure[67].

Many of the symptoms of the dust-related lung diseases are similar to those of asthma or chronic bronchitis and emphysema. Unless a physician probes for occupational history (which might be far back in the patient's life) or uses sophisticated diagnostic techniques, the occupational factor may be completely missed. But even when occupational exposure has been current, and workers from the same mine or mill were showing similar symptoms, the occupational factor has failed to be given due weight. The history of the diagnosis, the epidemiology, and the control (or lack of it) of byssinosis in the cotton mills makes sad reading. The same can be said of asbestosis in construction and other asbestos-using industries, and of black lung in the coal mines[68].

This history not only points to the need for much more vigorous control of occupational air pollution, but it contains a lesson for control of community air pollution as well. Correlation of present exposure and present illness may present a misleading picture. A case in point is a recent study of exposure to low concentrations of asbestos in shipyard workers whose job is the insulation of pipes[69]. The same shipyard had been investigated in 1945[70]. At that time, only three cases of asbestosis were found among 1074 pipe coverers, and these three had been exposed for 20 years or more. The investigators concluded that this was not a hazardous occupation. But at that time, less than ten percent of the workers had been exposed for more than ten years. The recent study found that signs and symptoms of asbestosis were not seen until a worker had at least ten years of exposure. With increasing years of exposure, the percentage of afflicted workers rose sharply.

The Poor and the Black

One thread that has appeared here and there throughout this discussion deserves to be picked up and followed a little farther. In U.S. and British industrial cities, the poor suffer the greatest exposure to air pollution. They also have a higher rate of illness and death from some of the diseases that have been associated with air pollution. Although social and environmental factors other than air pollution are involved, there is increasing evidence that air pollution is an important factor. For example, when persistent cough was found to be more prevalent among British children six to ten years old in lower socioeconomic classes than in upper classes, the difference was especially pronounced in cities with the most severe air pollution [40]. The relationship of air pollution to socioeconomic class and to childhood allergic disease in Buffalo has already been mentioned (page 133).

A series of related studies in the same city investigated total mortality and mortality from specific causes[71]. The area was divided according to suspended particulate pollution measured over two years at 21 stations. Particulates ranged from less than 80 micrograms per cubic meter of air to more than 135. The relationship between high air pollution and low economic status is shown by the fact that of more than 10,000 white males with incomes below $5007 whose deaths were studied, 2611 lived in the highest air pollution area, none in the lowest. At the other end of the scale, of more than 16,000 with incomes between $7431 and $11,792, 7680 lived in the lowest air pollution area and none in the highest.

The findings of this series of studies generally supported the existence of an association between total death rate and air pollution and a similar association of air pollution and death from chronic respiratory disease, stomach cancer, and prostatic cancer (but not lung cancer). That air pollution was indeed important in these death rates and was not being confused with some other factors related to socioeconomic status, was supported by the difference *within* economic groups whose air pollution exposure differed. For example, in the income range of $5175–$6004, which was represented in all four air pollution areas, the death rate from all causes among white men 50–69 years old was 50 percent higher in the highest air pollution area than in the lowest[71].

The Nashville studies attempted to isolate the air pollution effect by concentrating on middle-income mortality, since this class was represented at all levels of pollution. Other factors as well as air pollution appeared to be pushing up the death rate for respiratory disease in the low economic classes for, when various classes exposed to the same pollution level were compared, there were more deaths from this cause among the poor. Neither in Buffalo nor in Nashville was it possible to take occupational, smoking, or residential history into consideration. Only residence at time of death as related to current air pollution could be documented.

In the Nashville study, black and white death rates within areas of similar pollution could be compared only at moderate and high levels of pollution. Even in the middle class, where the study concentrated, the nonwhite population exposed to low pollution levels was very small. Death rates from lung and bronchial cancer in these moderate and high pollution areas were higher for whites; for other respiratory diseases, higher for nonwhites[72].

In the United States, a large proportion of the poor are black, and the Chicago study[49] previously discussed (pages 130–32) showed that similar patterns of mortality were revealed whether the population was divided along economic lines or along racial lines. Among the poor and the black, adult occupational exposures may be higher than in the general population, while children are more at risk in these areas from eating crumbling lead paint in old houses, adding to their body burdens of lead from inhalation. If these children are black, there is a possibility they may have the sickle-cell anemia that would render them particularly susceptible.

There is not enough information to be able to say how much greater is the risk to the poor and the black, but perhaps it is enough to know that they bear a disproportionate share of this burden, as of so many others.

Uncertain Threshold

There can be very little doubt that air pollution is bad for human health. There is, however, considerable doubt as to whether we know how much air pollution must be reduced to eliminate further suffering and death. Standards permitting specific concentrations of air pollutants could be justified if a threshold for each pollutant could be demonstrated—that is, a concentration below which there will be no ill effects.

For some pollutants, it is reasonable to assume a threshold: for example, in the case of a particular element essential to the body in very small amounts, but toxic in large amounts. Selenium has already been mentioned. Others are chromium, manganese, iron, cobalt, copper, zinc, and molybdenum, all of which have been measured in urban air[73]. At some level beyond the essential amount there must be a threshold below which the element is harmless but above which there is increasing risk of harm.

Other metals—lead, cadmium, nickel, mercury, beryllium, tin, antimony, and bismuth—are not needed by the body, do not do it any good in even small amounts, and are known to be poisonous. Although "threshold limits" have been established for occupational exposure to these metals, the assumption of a threshold is less reasonable in these cases. Standards for beryllium and mercury have been proposed, but there is as yet no standard for any of the other heavy metals. The only standard which applies to the rest is the standard for particulate matter, so that, for all practical purposes, the assumption is that 75 micrograms of particulate matter per cubic meter of air, no matter what its composition, will not harm people. When samples of particulate matter from ambient air are analyzed for heavy metal content, there is considerable difference among various cities (and a much greater difference between cities and nonurban areas) in regard to which metals are present and in what concentrations. In general, the heavy metals are a small, but increasing, fraction of the total[74]. In short, setting the standard for particulate matter at 75 micrograms per cubic meter of air does not take into consideration the possibility that there may be no threshold for some of the heavy metals or for the organic particles that may be carcinogenic.

Another reasonable basis for assuming a threshold is the presence of an excretory mechanism in the body, known to be effective for low concentrations, but not for high ones, or a repair mechanism effective within certain limits. Those who maintain that environmental lead is not a hazard, base this on the body's ability to excrete most of the lead ingested and inhaled. Lead that is inhaled (and not promptly exhaled) is more readily absorbed than lead that is taken into the stomach. It is generally agreed that a high exposure, either from ingestion (as in children who eat lead paint) or inhalation (as in people living beside a heavily traveled highway), may increase the lead level in the blood to the point that it can damage the red blood cells, the kidneys, and the nervous system (including the brain). There are also indications that lead slowly accumulates in the body tissues with aging. It is the contention of some investigators that there is a threshold below which the excretory mechanism is able to keep the body in lead balance; that is, additional lead taken in—up to this threshold—simply results in more lead being excreted. While it is true that when more lead is taken in, more is excreted, it is not clear that the second function keeps up with the first.

It is possible to demonstrate the existence of a threshold by a series of controlled dose-response studies, if lower doses produce less response, and at some point it is no longer possible to find any indication of harm. The requirements for such studies, however, are impossible to fulfill with human subjects and very difficult and expensive with laboratory animals. A very large number of subjects are required and study is necessary over a long period of time in order to watch for the effects of long-term response to a low-level dose. If genetic effects are suspected, the studies must continue for several generations.

In regard to the chemical pollutants which have been found to play a role in the production of cancer, a recent National Academy of Sciences report found that

> Neither epidemiologic nor experimental data are adequate to fix a safe dosage of any chemical carcinogen below which there will definitely be no tumorigenic response in humans. For these reasons, synthetic chemicals, such as food additives and pesticides that are known to be carcinogenic, must not be deliberately added to the environment. With regard to air pollutants, which contain a variety of defined and undefined carcinogens, the lowest possible exposure must always be insisted on[75].

Of all the known air pollutants, only a few trace metals mentioned are known to be essential to the human body. For no other air pollutants has a threshold been satisfactorily demonstrated. For radioactive pollutants, the evidence *against* the existence of a threshold is sufficiently strong that scientific and government bodies such as the United Nations Scientific Committee on the Effects of Atomic Radiation and the Federal Radiation Council have said that it is prudent to assume that any exposure to radiation, no matter how small, carries with it some risk. Biologic effects of ozone have been observed at such low levels as to suggest that for ozone, as for radiation, it is prudent to assume that there is no threshold.

The benefit/risk balance in radiation standards has never been clearly set forth by the standard-setters, and what public debate there has been has suffered from considerable confusion as a result. However, the public has repeatedly expressed itself on behalf of lower exposures—by pressing for an end to the atmospheric testing of nuclear weapons, by opposing the construction of nuclear reactors in hazardous situations or close to populated centers, and by moving for stricter standards at the state level.

In 1969 and 1970, the radiation standard for the general population was sharply attacked by John Gofman and Arthur Tamplin of the AEC's

Division of Biology and Medicine, who attempted to spell out the risk implicit in this standard. If everyone in the population actually received the allowable dose, Gofman and Tamplin estimated, on the basis of the available information about dose-response relationships, that there would be an annual increase in cancer and leukemia of 24,000 or more cases [76]. Actual exposures are presently much below the standard of 0.17 rad (unit of nuclear radiation) per person, per year, but we are only on the threshold of the nuclear age.

The accepted basis for exposure to radiation—appropriate for any pollutant with known harmful effects at high doses and no known threshold—is that there should be *no* exposure without good reason. Standards are supposed to be set on the basis of balancing the risk against the expected benefit of the exposure.

We know that present levels of air pollution are causing some harm to health and we cannot say with any confidence that reaching the present standards will end that harm, although it will certainly help. Additional information, which is being painstakingly assembled in many laboratories and in many community studies, will also help us achieve a more realistic approach to air pollution and health and will point the way to necessary revisions in the standards. But our knowledge will never be perfect and the crucial question is how to act on the basis of imperfect knowledge.

Environmental health in its broadest sense means the creation and maintenance of a healthy environment, an environment which will foster and protect the most vulnerable members of the population. Environmental health requires seeking in existing information, guides to maintaining an atmosphere most conducive to human health. On the basis of present knowledge, we are not likely to get firm answers to questions like: "How many additional cases of chronic bronchitis can be expected with every additional part per million of sulfur dioxide in the air?" "By what percent will hypertension increase with an increase of a half a microgram of cadmium per cubic meter of air?" But if we ask whether it is likely that reducing sulfur oxides, nitrogen oxides, ozone, and all the many forms of particles would result in improved health, the answer is an unequivocal "yes."

The danger in such a generalization is that it may be so vague as to become meaningless. It can lead to the principle of reducing air pollution "as much as possible," and the art of the possible has many interpreters. The value in such a generalization is that it can force us to recognize that, if we wish to protect the health of those most vulnerable to the damage of air pollution, we cannot afford to wait for proof that x number of people exposed to y amount of an *abc* mix of pollutants will produce z cases of asthma, emphysema, cancer, or birth defects. When there are so many unknowns, we are forced to gamble. Either we gamble with the lives and health of people, or we take a chance that we are paying more for control than we really need to in order to provide a genuine margin of safety, a truly healthy environment.

[1] Mary O. Amdur, "Toxicological Appraisal of Particulate Matter, Oxides of Sulfur, and Sulfuric Acid," *Journal of the Air Pollution Control Association,* September 1969, 19:638–46.

[2] Eric J. Cassell, "The Health Effects of Air Pollution and Their Implications for Control," *Law and Contemporary Problems,* spring 1968, 33:199.

[3] William W. Stalker, Richard C. Dickerson, and George D. Kramer, "Sampling Station and Time Requirements for Urban Air Pollution Surveys, Part III: Two- and Four-Hour Soiling Index" and "Part IV: Two- and Four-Hour Sulfur Dioxide and Summary of Other Pollutants," *Journal of the Air Pollution Control Association,* April 1962, 12:170–78 and August 1962, 12:361–75.

[4] Martin E. Pearlman et al., "Nitrogen Dioxide and Lower Respiratory Illness," *Pediatrics,* February 1971, 47:391–98. This study divided the city in three regions according to low, medium, and high nitrogen dioxide, and studied respiratory problems in school children. Both the relative proportion of children affected by bronchitis and the frequency of bronchial attacks were higher among children exposed for several years to the higher nitrogen dioxide levels.

[5] W. P. Shepard, *The Physician in Industry* (New York: McGraw-Hill Book Company, 1961), p. 2.

[6] L. Otis Emik et al., "Biological Effects of Urban Air Pollution," *Archives of Environmental Health,* November 1971, 23:335. During the time of this study, the oxidant average was 0.057 parts per million and the carbon monoxide average was 1.7 parts per million. The biological differences noted in the exposed, as compared to control, animals included a reduction in a pulmonary enzyme (alkaline phosphatase) in rats and a blood serum enzyme (glutamine oxalacetic transaminase) in rabbits; a higher prevalence of pneumonitis in mice; a shortened life for male mice. In one strain of mice, decreased running activity was highly correlated with oxidant concentration. An earlier article of the same title in the same journal by M. B. Gardner (March 1966, 12:305–13) reported more pneumonitis in mice living in ambient Los Angeles air than in those living in filtered air.

[7] L. G. Strandberg, "SO_2 Absorption in the Respiratory Tract," *Archives of Environmental Health,* August 1964, 9:160–64.

[8] Hans L. Falk, "Chemical Definitions of Inhalation Hazards," in *Inhalation Carcinogenesis, Proceedings of the Biology Division, Oak Ridge National Laboratory Conference in Gatlinburg, Tennessee, October 1969,* U.S. Atomic Energy Commission, Division of Technical Information, Oak Ridge, April 1970, pp. 13–24.

[9] Pennsylvania State University press release, November 23, 1971, reporting a study by Arian Zarkower.

[10] R. B. Chevalier, R. A. Crumholz, and J. C. Ross, "Reaction of Nonsmokers to Carbon Monoxide Inhalation," *Journal of the American Medical Association,* December 5, 1966, 198:135. Quoted by Stephen M. Ayres, Stanley Giannelli, Jr., and Hiltrud Mueller in "Carboxyhemoglobin and the Access to Oxygen: An Example of Human Counter-Evolution," presented at the AMA Air Pollution Medical Research Conference, New Orleans, Louisiana, October 1970.

(Other aspects of the discussion of carbon monoxide are taken from this paper by Ayres et al.)

[11] Julian McCaull, "Building a Shorter Life," *Environment,* September 1971, Vol. 13, No. 7, 2.

[12] Richard Ehrlich, "Effect of Nitrogen Dioxide on Resistance to Respiratory Infection," *Bacteriological Reviews,* 30:604–14. Also the same author with Mary C. Henry and James Fenters, "Influence of Nitrogen Dioxide on Resistance to Respiratory Infection," in *Inhalation Carcinogenesis,* September 1966, pp. 243–57 (see [8]).

[13] D. L. Coffin et al., "Influence of Ozone on Pulmonary Cells," *Archives of Environmental Health,* May 1968, 16:633.

[14] Eula Bingham et al., "Alveolar Macrophages: Reduced Number in Rats After Prolonged Inhalation of Lead Sesquioxide," *Science,* December 13, 1968, 162:1297–99.

[15] Studies by C. W. Labelle cited in *Air Quality Criteria for Particulate Matter,* U.S. Department of Health, Education and Welfare, NAPCA Pub. AP-49, January 1969, p. 123.

[16] I. J. Selikoff, "Asbestos," *Environment,* March 1969, Vol. 11, No. 2, 2–7, and Virginia Brodine, "A Special Burden," *Environment,* March 1971, Vol. 13, No. 2, 22–33.

[17] Lewis J. Cralley, "Inhalable Fibrous Material," *Pneumoconiosis, Proceedings of the International Conference, Johannesburg, 1969,* H. A. Shapiro (ed.) (Cape Town, South Africa: Oxford University Press, 1970), pp. 70 and 72.

[18] *Particulate Polycyclic Organic Matter,* report of the Committee on Biologic Effects of Atmospheric Pollutants. National Academy of Sciences, Washington, D.C. 1972, pp. 205–36.

[19] *Air Quality Criteria for Particulate Matter,* p. 22 (see [15]).

[20] P. Stocks, "On the Relations Between Atmospheric Pollution in Urban and Rural Localities and Mortality from Cancer, Bronchitis, and Pneumonia," *British Journal of Cancer,* September 1960, 14:397–418.

[21] R. M. Hagstrom, H. A. Sprague, and E. Landau, "The Nashville Air Pollution Study, Part VII: Mortality from Cancer in Relation to Air Pollution," *Archives of Environmental Health,* August 1967, 15:237–48. (For more information about the Nashville Study, see also [59] and [72].)

[22] Warren Winkelstein, Jr. and Seymour Kantor, "Stomach Cancer: Positive Association with Suspended Particulate Air Pollution," *Archives of Environmental Health,* April 1969, 18:544–47.

[23] Selikoff, "Asbestos," p. 5 (see [16]).

[24] Paul Kotin and Hans L. Falk, "Atmospheric Factors in Pathogenesis of Lung Cancer," *Advances in Cancer Research,* 1963, 7:457–514.

[25] Falk, "Chemical Definitions," p. 20 (see [8]). An exhaustive report on the polycyclic hydrocarbons has been issued by the National Academy of Sciences (see [18]).

[26] P. R. Peacock and J. B. Spence, "Incidence of Lung Tumors in LX Mice Exposed to (1) Free Radicals; (2) SO_2," *British Journal of Cancer*, September 1967, 21:606–18. Cited by Falk, p. 21 ([8]).

[27] Sidney Laskin, Marvin Kushner, and Robert T. Drew, "Studies in Pulmonary Carcinogenesis," in *Inhalation Carcinogenesis*, pp. 321–50 (see [8]).

[28] See James F. Crow, "Chemical Risk to Future Generations," *Scientist and Citizen* (now *Environment*), June–July 1968, Vol. 10, No. 5, 113–17, for a discussion of the importance and problems of monitoring environmental mutagens.

[29] Robert Shapiro, "Reactions of Uracil and Cytosine Derivatives with Sodium Bisulfite Deamination," *Journal of The American Chemical Society*, January 28, 1970, 98:422–24.

[30] Herbert E. Stokinger and D. L. Coffin, "Biologic Effects of Air Pollutants," in Arthur C. Stern (ed.) *Air Pollution*, second edition (New York: Academic Press, 1968), Vol. 1, pp. 460–65.

[31] J. A. Hildes, "Ecologic and Ethnic Adaptations," *Environmental Research*, 1969, 2:418.

[32] Stokinger and Coffin, "Biologic Effects of Air Pollutants," p. 457 (see [30]).

[33] René Dubos, "Adapting to Pollution," *Scientist and Citizen* (now *Environment*), Vol. 10, No. 1, 1–8.

[34] Crow, "Chemical Risk," p. 114 (see [28]).

[35] *Air Quality Criteria for Carbon Monoxide*, U.S. Department of Health, Education and Welfare, NAPCA Pub. AP-62, March 1970, pp. 9-8 and 9-9.

[36] W. P. D. Logan, "Mortality in the London Fog Incident, 1952," *The Lancet*, February 14, 1953, 336–38.

[37] S. S. Epstein et al., "The Carcinogenicity of Organic Particulate Pollutants in Urban Air After Administration of Trace Quantities to Neonatal Mice," *Nature*, 1966, 212:1305–07.

[38] T. Toyama, "Air Pollution and Its Health Effects in Japan," *Archives of Environmental Health*, January 1964, 8:153–73. Also J. E. Lunn et al., "Patterns of Respiratory Illness in Sheffield Infant School Children," *British Journal of Preventive and Social Medicine*, 1967, 21:7–16.

[39] Pearlman et al., "Nitrogen Dioxide," (see [4]).

[40] D. D. Reid, "The Beginnings of Bronchitis," *Proceedings of the Royal Society of Medicine*, 1969, 62:311.

[41] J. W. B. Douglas and R. E. Waller, "Air Pollution and Respiratory Infection in Children," *British Journal of Preventive and Social Medicine,* 1966, 20:1–8.

[42] Reid, "The Beginnings of Bronchitis," p. 311 (see [40]).

[43] Ibid., p. 312.

[44] The sex difference has been attributed to the difference in smoking habits and in occupational exposures. It has also been suggested that it may be related to the female's advantage over the male in resistance to serious bacterial infections. See Robert J. Schlegel and Joseph A. Bellanti, "Increased Susceptibility of Males to Infection," *The Lancet,* October 18, 1969, 2:826–27.

[45] A. S. Fairbairn and D. D. Reid, "Air Pollution and Other Local Factors in Respiratory Disease," *British Journal of Preventive and Social Medicine,* 1958, 12:94–103. Also, W. W. Holland and D. D. Reid, "The Urban Factor in Chronic Bronchitis," *The Lancet,* February 27, 1965, 1:445–48.

[46] Bertram W. Carnow et al., "The Chicago Air Pollution Study: Acute Illness and SO_2 Levels in Patients with Chronic Bronchopulmonary Disease," *Archives of Environmental Health,* May 1969, 18:768–76.

[47] Theodor D. Sterling, Seymour V. Pollack, and J. Weinham, "Measuring the Effect of Air Pollution on Urban Morbidity," *Archives of Environmental Health,* April 1969, 18:485–94. There are many other studies of air pollution and respiratory disease in England as well as in the United States. Additional references will be found in the federal criteria documents for sulfur oxides and particulates.

[48] Committee of the Royal College of Physicians of London on Smoking and Atmospheric Pollution, *Air Pollution and Health* (London: Pitman Medical and Scientific Publishing Co., Ltd., 1970), p. 68.

[49] Dr. Carnow [46] reported this study at the 1970 AMA Air Pollution Medical Research Conference in New Orleans. See Virginia Brodine, "A Special Burden," *Environment,* March 1971, Vol. 13, No. 2, 22.

[50] M. Glasser, L. Greenburg, and F. Field, "Mortality and Morbidity During a Period of High Levels of Air Pollution, New York, November 23 to 25, 1966," *Archives of Environmental Health,* December 1967, 15:684.

[51] L. D. Zeidberg, R. A. Prindle, and E. Landau, "The Nashville Air Pollution Study, Part I: SO_2 and Bronchial Asthma," *American Review of Respiratory Disease,* 1961, 84:489.

[52] H. H. Schrenk et al., "Air Pollution in Donora, Pennsylvania," Bulletin No. 306, Public Health Service, Division of Industrial Hygiene, 1949.

[53] L. S. Girsh et al., "A Study on the Epidemic of Asthma in Children in Philadelphia," *Journal of Allergy,* 1967, 39:347.

[54] Harry A. Sultz et al., "An Effect of Continued Exposure to Air Pollution on the Incidence of Chronic Childhood Allergic Disease," *American Journal of Public Health,* May 1970, 60:890.

[55] Stephen M. Ayres and Meta E. Buehler, "The Effects of Urban Air Pollution on Health," *Clinical Pharmacology and Therapeutics,* May–June 1970, 11:360.

[56] C. C. Hunter, Jr., J. A. Pierce, and J. B. LaBorde, "Alpha-1 Antitrypsin Deficiency, A Family Study," *Journal of the American Medical Association,* 1968, 205:23. See also, S. Makino et al., "Emphysema with Hereditary Alpha-1 Antitrypsin Deficiency Masquerading as Asthma," *Journal of Allergy,* 1970, 46:40.

[57] Charles Mittman et al., "Smoking and Chronic Obstructive Lung Disease in Alpha-1 Antitrypsin Deficiency," *Chest,* September 1971, 60:214–21.

[58] H. E. Stokinger, "Sanity in Research and Evaluation of Environmental Health," *Science,* November 12, 1971, 174:664. Stokinger suggests studies of asthmatics and people with a deficiency of leukocytic enzymes as well.

[59] L. D. Zeidberg, R. J. M. Horton, and E. Landau, "The Nashville Air Pollution Study, Part VI: Cardiovascular Disease Mortality in Relation to Air Pollution," *Archives of Environmental Health,* August 1967, 15:225–36. (For more information about the Nashville Study, see also [21] and [72].

[60] Alfred C. Hexter and J. R. Goldsmith, "Carbon Monoxide: Association of Community Air Pollution with Mortality," *Science,* April 16, 1971, 172:265–66.

[61] S. I. Cohen, M. Deane, and J. R. Goldsmith, "Carbon Monoxide and Myocardial Infarction," *Archives of Environmental Health,* April 1969, 19:510–17.

[62] Paul P. Craig and Edward Berlin, "The Air of Poverty," *Environment,* June 1971, Vol. 13, No. 5, 2–9.

[63] I. J. Selikoff, E. Cuyler Hammond, and J. Churg, "Mortality Experiences of Asbestos Insulation Workers," *Pneumoconiosis,* p. 183 (see [17]).

[64] Harold C. Hodge et al., paper presented at the AMA Air Pollution Medical Research Conference, New Orleans, Louisiana, October 1970.

[65] G. Saccomanno, "Radiation Exposure of Uranium Miners," *Radiation Standards for Uranium Miners, Hearings Before the Joint Committee on Atomic Energy, March 1969* (Washington, D.C.: U.S. Government Printing Office, 1969), pp. 303–09.

[66] Norton Nelson, "Some Biological Effects of Radiation in Relation to Other Environmental Agents," *Biological Implications of the Nuclear Age, Proceedings of a Symposium at Lawrence Radiation Laboratory, March 1968,* U.S. Atomic Energy Commission, Division of Technical Information, Oak Ridge, Tennessee, 1969, pp. 223–29.

[67] R. Doll et al., "Mortality of Gasworkers with Special Reference to Cancer of the Lung and Bladder, Chronic Bronchitis and Pneumoconiosis," *British Journal of Industrial Medicine,* 1965, 22:1–12. (Cited in *Particulate Polycyclic Organic Matter* [18], p. 202.)

[68] *Occupational Health and Safety Act, 1970, Hearings Before the Subcommittee on Labor and Public Welfare, U.S. Senate, 91st Congress* (Washington, D.C.: U.S. Government Printing Office, 1970). Also, 1969 hearings before the same subcommittee on *Coal Mining Health and Safety.*

[69] Raymond L. H. Murphy, Jr. et al., "Effects of Low Concentrations of Asbestos," *New England Journal of Medicine,* December 2, 1971, 285:1271–78.

[70] W. E. Fleischer et al., "A Health Survey of Pipe Covering Operations in Constructing Naval Vessels," *Journal of Industrial Hygiene and Toxicology,* 1946, 28:9–16.

[71] Warren Winkelstein, Jr. et al., "The Relationship of Air Pollution and Economic Status to Total Mortality and Selected Respiratory System Mortality in Men: I Suspended Particulates," *Archives of Environmental Health,* 1967, 14:162–71. (Also see Warren Winkelstein, Jr. and Seymour Kantor, "Prostatic Cancer: Relationship to Suspended Particulate Air Pollution," *American Journal of Public Health,* July 1969, 59:1134–38 and [22].)

[72] L. D. Zeidberg, Robert J. M. Horton, and E. Landau, "The Nashville Air Pollution Study Part V: Mortality from Diseases of the Respiratory System in Relation to Air Pollution," *Archives of Environmental Health,* 1967, 15:214–24. (For more information about the Nashville Study, see also [21] and [59].)

[73] Henry A. Schroeder, "Metals in the Air," *Environment,* October 1971, Vol. 13, No. 8, 18.

[74] Gary D. Carlson and Wayne E. Black, "Determination of Trace Quantities of Metals from Filtered Air Samples by Atomic Absorption Spectroscopy," presented at the 63rd Annual Meeting of the Air Pollution Control Association, St. Louis, Missouri, June 1970. Summarizing work done on iron, lead, zinc, copper, manganese, and cadmium in St. Louis, the authors say:

> . . . The six trace metals which were examined show marked increases in nearly all cases for 1969, compared to 1963 and 1964. Although the finite numbers of the concentrations may be of little importance at the present time, the fact that there is a trend of increasing levels in such a few short years is of primary importance.

[75] *Particulate Polycyclic Organic Matter,* p. 93 (see [18]).

[76] John W. Gofman and Arthur R. Tamplin, "Radiation: The Invisible Casualties," *Environment,* April 1970, Vol. 12, No. 3, 12.

Reducing the Burden

Between the individual and his natural environment stands another complex system—the economic, political, and social system of the country in which he lives. This is the medium through which most of our relationships with our natural environment are conducted, and it is here that reorganization of human activity is required to reduce the burden of contaminants in the atmosphere and in our own bodies.

Those individuals who enjoy spending their leisure time in a remote mountain wilderness and those who prefer to be where the city lights are brightest may feel and think very differently about nature. The household that recycles its newspapers, rather than burning them, and uses a bicycle or public transportation, rather than a second car, may be expressing a greater concern about air pollution and contributing a mite less to its city's total. Nevertheless, if we actually reduce the burden, it will not be through the choices individuals make in their private lives, but through the choices they make collectively to adapt or change an economic system which now is heavily dependent upon the largely uncontrolled exploitation of nature and a political system whose decision-making processes were not developed to cope with environmental problems.

Current Efforts to Reduce the Burden

Air pollution control programs scarcely went beyond a few smoke control ordinances and nuisance laws until 1947 when Los Angeles launched its long fight against smog. It was more than a decade later that this city was

followed haltingly and unevenly by other cities. In the sixties, the federal government increasingly provided education, research, technical assistance, monitoring, and finally, in 1967, some steps toward control. Without those efforts, the situation would undoubtedly be worse than it is today; nevertheless, although some communities succeeded in holding the line or even making some gains, overall the air was almost certainly worse at the end of the decade than it was at the beginning.

The seventies began with new federal legislation (an overhauling of the Clean Air Act) and with uniform ambient air quality standards for six major pollutants: sulfur dioxide, particulates, carbon monoxide, photochemical oxidants, nitrogen oxides, and hydrocarbons. Attainment of the air quality specified by these standards is left to the states, but they are required to submit their plans to the Federal Office of Air Programs for approval. Once the plans are approved, the states then have three to five years to reach the goals set by the federal standards.

When the primary standards based on health effects have been attained, air pollution control regions are expected to move on to more restrictive secondary standards (based on environmental effects) for sulfur dioxide and particulates. Those regions where air quality is now as good as, or better than, the primary standards are to move directly to secondary standards. Outlying areas where the air quality is even better than secondary standards should not allow "significant deterioration" of this quality. Emergency provisions are required to prevent "excessive build-up" of pollution levels during periods of stagnant air when pollutants cannot disperse.

The standards for ambient air are reinforced by some restrictions on stack emissions. Standards are for new installations in a number of categories, regardless of location. There will be uniform national emission standards for a few of the contaminants that are hazardous even in small amounts: The proposed beryllium, mercury, and asbestos standards are in this category, and others may be added. Emission standards for automobiles have been specified: New cars in 1975 and thereafter are to emit 90 percent less hydrocarbons and carbon monoxide than in 1970, and 90 percent less nitrogen oxides than in 1971.

All of this makes it sound as if we should have the air cleaned up within the next few years, but that is far from the case. While there is some hope that automotive emissions will drop substantially in coming years, in the long run the present approach to air pollution control will *not reduce the total burden of pollutants* from stationary sources: industry, power generation, agriculture, space heating, and waste disposal. Gains in controlling emissions from each single source will be offset by the increase in the size and number of sources. Moreover, it is still quite uncertain just what the control devices on automobiles will achieve. In any case, these devices will have their maximum effect on pollution reduction about 1990 and emissions from this source will then begin to climb again if the number of automobiles on the road continues to increase. Either we will fail to meet the goal of improving air quality in the cities over the long haul or we will fail to meet the goal of protecting air outside the cities from deterioration, or both.

Fossil-Fuel Burning

A very big part of our air pollution problem arises from the burning of fossil fuels for the generation of electric power, for industrial processes, and for space heating. Federal air pollution control engineers have summarized the situation for the remaining years of this century:

> If we assume that the work we are doing now to develop and apply control methods is successful, overall amounts of sulfur oxides and nitrogen oxide emissions can be held to relatively modest *increases over present levels.* . . . SO_2 emissions from the power industry are expected to increase *with controls* by an estimated two to three times by the year 2000.[1]

Nitrogen oxide control techniques are behind those for sulfur oxides and the control engineers did not venture any estimates of what the nitrogen oxide emission increases from fossil fuels would be *with* controls. Without controls, however, they estimate that these emissions will be about 11 million tons in 1970 and about 25 million tons in the year 2000. This is a conservative estimate, as the engineers themselves point out, because their estimates are based on emissions of older boilers, while "modern boilers operate at higher temperatures and therefore are expected to emit substantially higher concentrations of NO_x."[2]

Achievement of the new standard for sulfur dioxide is expected through a change from high-sulfur to low-sulfur fuels in those cities where the sulfur pollution is most severe. However, emissions from burning low-sulfur coal are apt to contain more particulates than emissions from high-sulfur coal. Gas is doubly desirable because it produces less sulfur *and* less particulates. The use of low-sulfur fuel has limitations when we look to the future, however, as supplies are not unlimited.

Another major pollutant from fossil-fuel burning—particulates—*is* expected to decrease. Techniques now available for removing large particles and some smaller ones from stacks have cut particulates from fuel burning to the present annual 5.95 million tons. Wider use and further improvements of removal techniques may reduce particulates to about one-third of this amount by 2000. What will happen after that depends on how much we rely on fossil fuel and how much on nuclear power or other energy sources for the generation of electricity. If fossil-fuel combustion continues to increase, particulates would begin to climb again after the turn of the century.

Unfortunately, the controls that remove particles down to one micron in diameter quite efficiently take out only a small percentage of the finer particles which affect visibility, weather, climate, and health. The graph below projects the inexorable rise in fine particles below one micron in

Fine particles below 1.0 micron in diameter are the most difficult to control, remain longest in the air, and penetrate most easily deep into the lung. The projected rise in fine particle emissions shown here is from coal-fired utility boilers only.

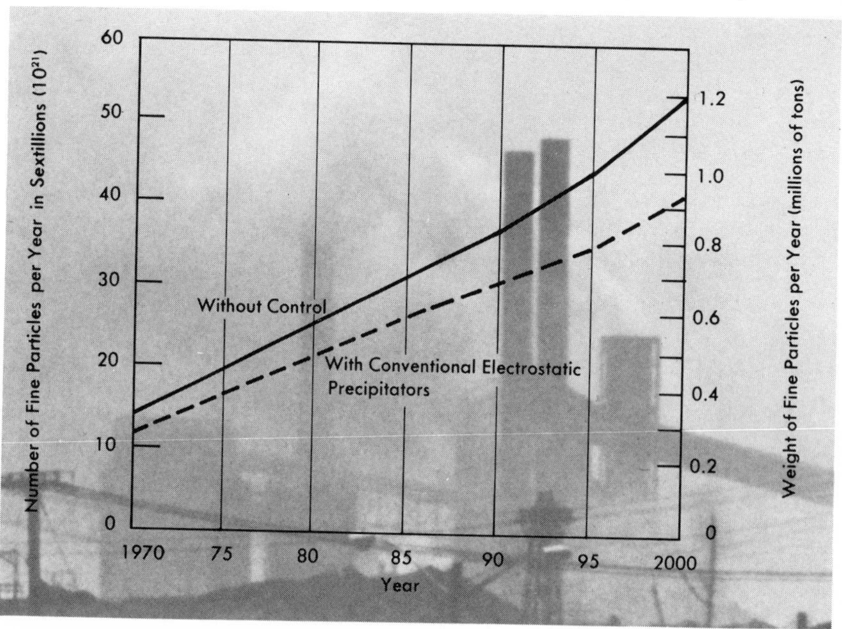

diameter. This rise may be as conservative an estimate as that projected for the nitrogen oxide emissions from fuel burning, and for the same reason: Higher temperatures in the power plants of the future would be expected to produce a larger proportion of finer particles[3].

Although fuel burning is the single largest source of particulates, it accounts for only about one-third of the total, as shown in the table* of major industrial sources of particulate pollutants. This 1970 Midwest Research Institute study concluded that particulate emissions could be reduced to 11 million tons per year by 1980 by the installation of cur-

Source		Emissions in Tons per Year
Fuel Combustion		
Electric Utilities	3,169,000	5,953,000
Industrial Use	2,784,000	
Crushed Stone, Sand, and Gravel		4,600,000
Agricultural and Related Operations		1,768,000
Iron and Steel (includes iron foundries)		1,564,000
Cement		934,000
Forest Products		666,000
Nonferrous Metals		
Copper	302,000	
Aluminum	200,000	
Zinc	62,000	602,000
Lead	38,000	
Lime		573,000
Clay		468,000
Fertilizer and Phosphate Rock		327,000
Asphalt		218,000
Ferroalloys		160,000
Coal Cleaning		94,000
Carbon Black		93,000
Petroleum Refining		45,000
Acids (sulfuric and phosphoric)		16,000
TOTAL**		18,081,000

* Excerpted from "Particulate Air Pollution in the United States," by A. Eugene Vandegrift et al., *Journal of the Air Pollution Control Association*, June 1971, 21:324–27.

** This total does not include particulates from forestry slash burning and agricultural field burning, previously estimated by the authors at 6,000,000 and 1,000,000 tons per year respectively.

rently available control devices on all sources and some improvements in control devices[4].

Pollution from the Automobile

The tailpipe of the automobile emits fine particles at a level where they very easily reach the human lung, yet the present emission controls for automobiles do not affect these particles. It was reported in 1969 that "an auto equipped with the so-called antipollution devices is as effective a source . . . as a car without such a control mechanism. If anything, it appears to produce even more. . . ."[5]

The new requirements for automobile pollution control effective in 1975 do not specify any improvement in particulate control, and there is no reason to think they will achieve any, except indirectly through changes in gasoline composition. The elimination of lead as an additive, for example, would be an important contribution to air quality, both in relation to health and in relation to inadvertent weather modification. Lead may be phased out because it interferes with the functioning of pollution control devices, but there is some concern that it may be replaced by other undesirable additives. The Clean Air Act presently requires the registration of new additives and permits the Environmental Protection Agency to control them, but an inquiry by *Environment* magazine in the spring of 1970 about a new additive revealed that:

> . . . no one has any information about the effects of the new additive TAP on health; no one knows whether it is more or less dangerous than the TMP which it has already replaced and the tetraethyl lead which it will be replacing. . . . As things look now, it will be at least two years before the federal government is even in a position to ask questions about the health effects of TAP, which is meanwhile being added in considerable quantities to the environment. . . .[6]

The technology for producing high-octane gasoline that would need no antiknock additive has been available for 12 years, but this would require refinery conversion estimated at over two billion dollars. However, the petroleum industry would save this amount in five years, since it now pays about $450 to $470 million annually for tetraethyl lead. Moreover, there would be an added benefit because changes in the refining procedure would free many thousands of barrels of propane, butane, and light naphtha for sale to the petrochemical industry. Nevertheless, in the spring of 1971, *Chemical & Engineering News* reported that "most refineries have yet to make a firm commitment to leadfree gasoline."[7] At any rate, they have not made the commitment to refinery conversion, and may be hoping to solve the problem with a substitute additive.

Although lead has been most widely discussed, it is only one of many gasoline additives. Manganese and nickel are also used as antiknock additives. Other additives are ethylene dichloride, phosphorus and boron compounds, alkylated phenols or other antioxidants, metal deactivators, antirust agents, anti-icing agents such as alcohol or phosphates, and lubricants[8]. Many of these contribute to the particles in automobile emissions. The metallic additives such as boron and nickel are especially suspect in terms of producing atmospheric effects similar to those of lead. In addition, they make it more difficult for the engine to burn the fuel completely, and therefore add to the amount of unburned hydrocarbons coming out the tailpipe.

The requirement of 90 percent reduction by 1975 in emissions from new automobiles will not eliminate 90 percent of the present air pollution attributed to the automobile. There will still be many pre-1975 cars on the road; the number of cars and miles traveled per vehicle and the concentration of cars in cities and their environs are expected to continue rising, and the efficiency of the control devices is expected to deteriorate with age. Urban emissions of hydrocarbons and carbon monoxide are now dropping and nitrogen oxides are expected to begin a downward trend before 1975, but when they reach their lowest point, in about 1990, the reductions,

Projected national urban emissions from motor vehicles. Hydrocarbon and carbon monoxide pollution from automobiles reached an all-time high in 1967, and then began to drop with the introduction of control devices which increased the combustion of the fuel and left less residue to be emitted. However, these same control devices achieved the desired effect by increasing the engine's air intake, *thus increasing nitrogen oxide pollution.* This graph shows the estimated reduction in automobile emissions *if standards set for present and future cars are met.* Maximum improvement will be reached in about 1990, and then, if the number of cars on the road is still increasing, automobile pollution will begin to rise again.

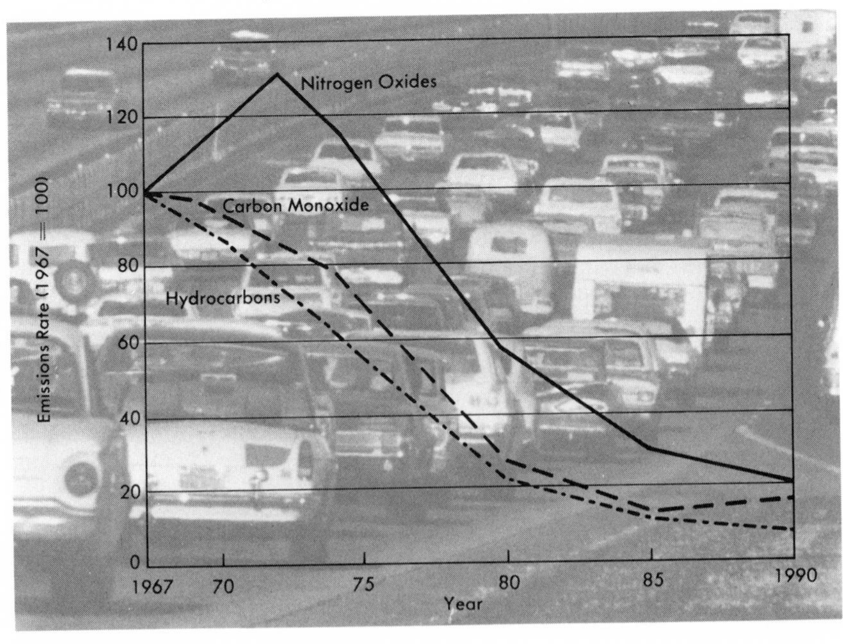

according to EPA projections, will be about 93.5 percent for hydrocarbons and about 85 percent for carbon monoxide from their respective highs in 1967 and 80 percent for nitrogen oxides from their high in 1972 or 1973[9]. Furthermore, these projections are very optimistic. They assume that between now and 1975 the new models will meet present standards, although tests so far have shown that most cars are not doing so[10]; they also assume that the 1975 models will be able to meet the more restrictive standards set for that year, although spokesmen for the automobile industry are still saying this is impossible.

What will happen to the complex chemistry of the air as the ratios of one pollutant to another are changed is unclear. Reduction in hydrocarbons and carbon monoxide from the early automotive control devices was achieved at the expense of a rise in nitrogen oxides. Also, the changing ratio of nitrogen oxides and hydrocarbons may have been responsible for another effect noted in Los Angeles between 1962 and 1967: Although the daily photochemical oxidant peaks were lower, the overall oxidant exposure was greater[11].

Other Pollutants

The expectation of increased use of nuclear power for the generation of electricity (see following graph) implies a rising curve for radioactive air pollution. Nuclear power plants, with their associated fuel-processing facilities, will be producing 470 million curies of krypton 85 by 2000[12]* How much of this is released to the atmosphere depends upon what controls are developed and what standards are set. Environmental radiation exposure, long in the hands of the Atomic Energy Commission and the Federal Radiation Council, is now the province of the EPA, although that agency has not yet taken action in the radiation field. The AEC is responsible for reactor design and operating criteria affecting radioactive emissions to air and water, but these criteria are subject to EPA approval. After coming under severe attack in the state of Minnesota[13] and from John Gofman and Arthur Tamplin of the AEC's own Lawrence Radiation Laboratory, the AEC recently issued new guidelines on emissions from water-cooled nuclear reactors. Reactors that presently exceed the guidelines have three years to improve, and even then will apparently be dealt with leniently. The guidelines do not apply to fuel-processing plants; the only one now operating emitted one million curies of krypton 85 in 1970[14]. Even if 99 percent of the iodine 129 produced in the reactors of the future is removed before the effluent reaches the environment, the breeder reactor program, as presently planned, will increase the present problem substantially. It is estimated that there could be 20 thousand curies of iodine 129 in the environment by 2060, presumably concentrated in the vicinity of,

*A curie is a unit of radioactivity equivalent to the amount given off by one gram of radium. By the time radioisotopes reach humans, their radioactivity is usually measured in picocuries (millionths of a millionth of a curie). The Radiation Protection Guideline for iodine 131 would be reached if the diet averaged 100 picocuries per day.

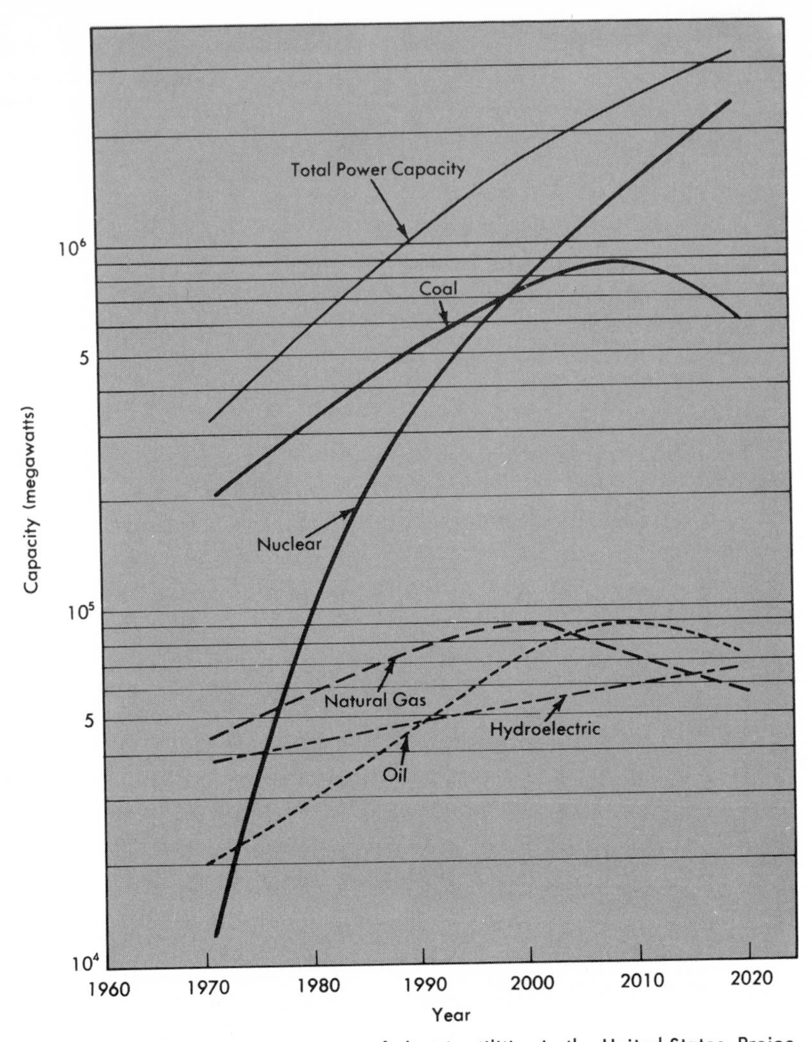

Projected power generating capacity of electric utilities in the United States. Projections for increased emissions of sulfur dioxide, nitrogen dioxide, fine particles, and radioactive contamination are based on the assumption that industry in general and electric utilities in particular will continue to grow as they have in the recent past. This projection of the growth of power generating capacity for the half-century 1970–2020 assumes success in the present effort to develop and operate fast-breeder nuclear reactors.

and downwind from, breeder reactors and fuel-reprocessing plants, and this amount would be constantly increasing[15].

There are as yet no standards for some of the known pollutants such as the chlorinated hydrocarbons, and new pollutants can be expected to appear in the atmosphere from the manufacture, use, and disposal of new products, altering the composition of air pollution. New pollutants may be discovered, investigated, and finally controlled, only to be replaced by some other contaminant. Past experience and present standard-setting procedures suggest that control of a new or previously uncontrolled pollutant will lag years behind its introduction into the atmosphere. As one scientist put it after years of air pollution research, "The statutory control of air pollution by pursuing air pollutants one by one as evidence accumulates seems clearly inadequate to a technology producing new pollutants or sources of pollution at an almost geometric rate."[16] Under the present law, a pollutant cannot be declared hazardous and its emission cannot be prohibited or controlled unless it will (in the judgment of the Environmental Protection Agency administrator) "cause, or contribute to, an increase in mortality or an increase in serious irreversible, or incapacitating reversible, illness."[17] New standards for pollutants not defined as hazardous are unlikely to be set until completion of the lengthy process of preparing a criteria document with proof that adverse health effects are produced by given concentrations of specific pollutants. (The language of the law is somewhat contradictory and confusing in this regard, but so far EPA has shown no disposition to stretch it.)

Control Strategies

The various estimates of future air pollution all assume that we *will* do what we *can* do to control it. Whether this proves to be true and, if so, the speed at which control is accomplished, depend on many things besides available technology: the effectiveness of regulatory strategies and staffs, whether industry moves willingly or must be forced every step of the way, and how much the public wants action and is willing to give political and financial support to local and federal control programs.

The uniform national emission standards for new installations in a number of industries are intended to ensure that, as old plants are phased out over the years, more and more—and, finally, all—plants will have the *best available, economically feasible* control technology. This will take 30 or 40 years.

Meanwhile, the present strategy is still reliance on ambient standards and the "air quality management" concept which does *not* require that everything that can be done for control actually be done. Essentially, ambient standards allow the air to get as dirty as we can stand it; the minimum tends to become the maximum—clean air tends to get dirtier at least to the level of the secondary standards. The principle of ambient standards based on health effects accepts the notion that anyone has the right to use the air for waste disposal as long as he does not endanger the health of others. One air pollution control official recently characterized the air quality management approach as obsolete in the face of public demands to bring pollution down *below* levels known to be dangerous to health[18].

Air quality management requires plans for reduction of pollution which take into account both emissions and the so-called self-cleansing properties of the air. It is a continuation of the concept of the sixties—the "optimum use of our air resources" for waste as well as for breathing. It is addressed solely to local situations, with no consideration of the national and international transport of pollutants or of the danger of inadvertent weather and climatic change. The best available technology need not be used, but only whatever technology will suffice to bring ambient air quality to the level prescribed by the standards.

Considerable emission monitoring and meteorological data is required for this strategy, and it tends to be obscured from the public in a cloud of technical complexities. This is particularly true if a region's plan utilizes a "diffusion model": a computer simulation of an area's atmosphere—pollutants emitted, their diffusion from their sources, and the impact of each point or area source on the area's air quality. Diffusion models work best in level terrain and cannot include the chemistry of the atmosphere. They can achieve, at best, very rough approximations of reality and are particularly unsuited to cities where photochemical smog is a problem. (A different approach to modeling which simulates the development of photochemical smog is under development, but is not yet being used in control planning.)

A much simpler proportional model can be used which, especially at this stage of diffusion modeling, has many advantages. Once the proportion of clean-up necessary for each pollutant has been determined, the plan for achieving it can be worked out, with consideration given to the effectiveness and cost of available control technology for the various sources[19]. Weather conditions and atmospheric chemistry, over which man has no control, do not figure in proportional models except as they are reflected in the ambient air measurements that provide the original baseline and mark progress toward the goal.

One of the problems that has surfaced as states have submitted their implementation plans arises from the number of days when air pollution can be expected to be much higher than usual because of unfavorable meteorological conditions. In order to prevent episodes endangering health, there are 24-hour standards not to be exceeded more than once a year. However, many air quality control regions will need to meet goals *below* the annual standard in order to ensure that they will not have as many as ten to 20 days in which the 24-hour standard is exceeded[20], or they will have to be prepared to put emergency controls into effect on those days.

An important aspect of any control plan is the location of ambient air monitoring stations. If a city has only one or a few stations, they tend to be placed near the area of heaviest pollution. When new stations are added, they are often population- rather than pollution-oriented and, if the data from the various stations are averaged, it will appear that a dramatic improvement has occurred. If most of its monitoring stations record the heaviest pollution, a community will be in the best position not only to watchdog pollution sources, but also to protect the health of the most heavily exposed residents and to assure that dangerous emergencies will not develop. However, in practice, the sites of available city- or state-owned property, an interest in getting a representative view of the community's ambient air, the desire to avoid the gloomy picture that might result from source-oriented monitoring, and the influence of polluters often lead to other decisions about siting.

At present, there are only about 14,000 monitoring devices in operation in the United States. Half of these are the simplest type—dustfall jars, sticky paper to catch windblown particles, and sulfation "candles" or "saucers."[21] This means that of the money allocated to air pollution control in the next few years, a considerable portion will be spent to develop surveillance networks, and it must be remembered that ambient air monitoring is still a very inexact technology.

As long as pollution is being emitted on a scale that threatens a disaster when weather conditions are unfavorable, continuous monitoring is required for warning and control of episodes. As long as some part of the population may be exposed to unacceptable pollution levels because of its proximity to sources or because of peculiarities of weather or terrain, monitoring is required for its protection. Once emissions are reduced to levels where there is an adequate margin of safety for unfavorable weather or poorly located residential areas, population-oriented monitoring becomes unnecessary. Thus, the more adequate a community's control program, the less it will need to be monitored; eventually only the watchdog function to ensure that controls continue to function properly would be necessary and this could be done at least partially by portable monitoring equipment.

The air quality management concept has been recognized as totally unsuitable for dealing with mobile sources. The car that appears to need no controls to protect air quality on a lightly traveled road or in a small town needs stringent controls to prevent it, in company with thousands of its kind, from emitting unacceptable amounts of carbon monoxide, nitrogen oxides, and hydrocarbons on a city street or heavily traveled highway. The automobile market is nationwide and, from a manufacturer's standpoint, a variety of standards would be impractical. After California led the way, uniform national emission standards were set by the federal government.

Other mobile sources will also have to be dealt with in the same way, and the legal basis for this has been laid in the Clean Air Act's 1970 amendments for uniform emission standards for vessels, aircraft, commercial vehicles, and new noncommercial vehicles.

The control of automobile pollution by tack-on emission devices and minor changes in engineering can be seen to be inadequate even before these devices and changes have been perfected. When the EPA announced the ambient standards, its administrator, William D. Ruckelshaus, said that "to meet the legal deadline for carbon monoxide . . . some cities may have to require drastic changes in their commuting habits."[22] In other words, these cities may have to control traffic patterns by staggering working hours, prohibiting traffic in certain sections of the city, and increasing the share of commuting handled by mass transit. These are social changes of considerable significance in cities where urban sprawl developed on the basis of uncontrolled use of the private automobile.

The Los Angeles Air Pollution Control District, the nation's most successful in controlling *stationary* sources, did *not* rely on the clumsy air quality management approach, but worked directly toward maximum abatement with available technology. The role of an aroused and educated public was crucial to the success of this effort[23].

Industry's record in Los Angeles, as in other parts of the country, has been generally one of an effort to delay and limit air pollution control. Industry spokesmen have claimed that the health effects have been unproved or exaggerated, standards and regulations too restrictive, and economic effects inadequately considered. Public hearings at every level from county to Congress are replete with these arguments, while efforts to abate pollution at every governmental level have been stalled by delaying tactics and court action. There are good dollars-and-cents reasons for this. Benjamin Linsky, San Francisco's first air pollution control officer and now a professor of air pollution control engineering, has told Congress that each year a public utility defers the installation of a million dollars worth of air pollution control equipment, the utility is ahead about $200,000 to $300,000[24].

Industry spokesmen now declare their devotion to the cause of clean air, but there is seldom a specific instance where a more comprehensive law, a more restrictive standard, a stiffer regulation, or a new control measure has been proposed that industry has not opposed it, recommended greater leniency, or urged lengthening the period before it became effective. Air pollution control is costly and industry wants to keep costs down in order to maintain a profitable operation, for profit is the key to our economic system.

As standards apply to additional pollutants and become more restrictive, the control technology generally becomes more expensive per unit of pollution controlled. Moreover, even as one control technique becomes available, an improved method is often being developed and a third may be in the research stage. This provides a ready-made case against installing today's technology: Either yesterday's is adequate, while today's will accomplish little more at great cost, or tomorrow's is just over the horizon and therefore today's would simply be a waste of money. These are all real problems and require careful consideration by legislators and regulators and technical and cost analyses by industrialists. What they add up to, however, is the conclusion that it would be unrealistic to expect any real change from industry's past approach: Industry will probably have to be forced to move every step of the way toward increased control.

While much of industry's past response to air pollution control can be found in the records of public hearings, these provide only a small part of the whole story. Spokesmen for industry frequently have an opportunity to work out compromises with control officials before legislation or regulatory measures are aired in public[25].

A mechanism for institutionalizing industry's participation in pollution control legislation and regulation on the federal level was instituted when President Nixon established the National Industrial Pollution Control Council. Attached to the Department of Commerce, one of the Council's mandates is to advise on plans and actions of Federal, State, and Local agencies involving environmental quality policies affecting industry[26].

It is clear that the public needs spokesmen who understand the technical aspects of air pollution and its control, and representatives who are conversant with the way the community's—and the nation's—legislative and regulatory bodies work. Reports like those from the Stanford Workshop[25] and the *Task Force Report on Air Pollution* by Ralph Nader's Center for the Study of Responsive Law[27] are real services to the public. Independent scientists acting individually, as consultants to citizen groups, and in science information committees associated with the Scientists' Institute for Public Information, have frequently spoken to the technical aspects of air pollution issues. However, these issues tend to be narrowly defined within the limits of what can be accomplished with currently available, economically feasible, tack-on controls. As long as this is the case, proponents of stricter air pollution control could win all the battles and improve the air quality in their own communities, but still lose the war to reduce the burden in the nation as a whole.

A case in point is electric power generation in Los Angeles. A series of regulations reduced particulates and substituted low- for high-sulfur fuel, making a major contribution to the reduction of emissions from this source. However, with power generation growing, even the reduced emissions of sulfur and particulates threatened to increase to an unacceptable level, and no technology was available that could adequately control nitrogen oxides. Finally, Los Angeles County decided against locating any further oil- or coal-burning power plants within its borders. This was one of the reasons for the construction of the Four Corners Plant close to where Utah, Colorado, Arizona, and New Mexico meet, to be followed by five additional giant plants planned by the same Western Energy Supply and Transmission Associates, a consortium of 23 utilities[28]. The present emission of roughly 400 tons of particulates per day from the five operating units of the Four Corners Plant can be compared with the daily average emitted in the entire New York metropolitan region, estimated at 633 tons per day in 1966. Sulfur dioxide emissions from the five units total about 300 tons per day.

Global Pollution

Neither waste heat nor carbon dioxide production will be affected by our present approach to control, and fine particles in the atmosphere will continue to increase. Some other interferences with natural cycles will be reduced (carbon monoxide) or their rate of increase slowed (sulfur and nitrogen), while others will scarcely be touched (synthetic organics). Moreover, as other nations become more industrialized, the global burden will

The effort to prevent further deterioration of city air has been one factor in the trend toward building electric generating plants near energy sources. Here giant transmission lines march off into the distance, carrying power to the cities from the Four Corners plant, while the southwestern sky is filled with clouds of pollution.

grow. A very different approach to the environment is found in the developing countries than in the highly industrialized countries. The principal environmental problems in the developing countries are hunger, disease, and poverty, and leaders fear that money spent on pollution control will add to industrialization costs and impede development. As Prime Minister Indira Gandhi of India said to the United Nations Conference on the Human Environment in Stockholm, June 5–16, 1972, "How can we speak to those who live in villages and in slums about keeping the oceans, the rivers, and the air clean when their own lives are contaminated at the source?"[29]

Since the United States achieved its present power and affluence not only by an uncontrolled exploitation of its own environment, but by extracting raw materials, changing agricultural patterns, and introducing polluting technologies in the less developed nations, we are in a poor position to object when these countries dump waste into the atmosphere. Unless this country moves to reverse the trend toward ever greater pollution, rather than simply slowing its rate of increase, we can hardly expect anything better from other nations nor should we be surprised at complaints from abroad as more and more nations realize what we have done and are continuing to do to the atmosphere we all share.

The best we can hope for if we continue along our present path is that the current high concentrations of pollutants in the cities will be lowered and the deterioration of air quality in the suburbs and in rural and wilderness regions will be slowed in some places. If the American people choose to settle for this path, they should do so with their eyes open—recognizing the nature of the gamble with their own health and the health of succeeding generations, recognizing the nature of the gamble with the life-sustaining qualities of the earth-atmosphere system. There is another choice.

Another Look at the Future

Dirtier air is not inevitable. It only seems so because we assume that our economy, and especially its energy consumption, will continue to grow at the same rate as in the recent past and that we will utilize the same or slightly improved techniques to control the pollution that results. When we take our present way of doing things as our starting point and assume that the control of air pollution must disturb the economy and its supporting technologies as little as possible, we focus on economically feasible, tack-on control devices.

If, on the other hand, we want a substantial reduction in the burden of air pollution, a much more fundamental approach is required. Instead of adapting nature to *man* in just slightly different ways, we need to find an alternative course in basically new ways to adapt *man* to nature, for, as Barry Commoner shows in *The Closing Circle,* our pollution problems arise from a "great fault in the life of man in the ecosphere."

> We have broken out of the circle of life, converting its endless cycles into man-made, linear events: Oil is taken from the ground, distilled into fuel, burned in an engine, converted thereby into noxious fumes which are emitted into the air. At the end of the line is smog.[30]

Learning to recognize that all life is a series of cycles and that the human species has no choice but to adapt to them if it is to survive is a shocking change in our perception of reality. It is as hard for this generation to grasp as the concept of a round earth was for an earlier generation. (And just as there is still a Flat Earth Society, we can expect that some people will fail to grasp the concept of the cyclical nature of life hundreds of years hence.) The hazards of our present course are becoming obvious, with the extraction of materials from the earth, the use of energy, and the production of goods all on an unending rise. This is burning our candle at both ends with a vengeance, for the inevitable outcome is depletion at one end and destruction of air, water, and the life-sustaining qualities of land at the other. It is easier to see these hazards than to chart a new, cyclical course for human events to replace the linear one. Discussion of such questions is only just beginning.

Each step of the way will require social, as well as scientific, decisions. Within the overall concept, we will need new approaches to *determining goals and standards, stimulating scientific advance and technological change, using resources more efficiently, revising our outlook on the growth rate, and making the economy more responsive to environmental concerns.*

Whether such basic changes in economic and political decision making can be justified for the sake of air pollution alone might be questioned, but the question is really academic. The same kind of changes are being put on the agenda by problems other than dirty air: depletion of underground resources, water pollution, soil depletion, and the unwieldy growth of cities and the deterioration of their manmade environments. Moreover, these environmental issues are intimately interconnected with the issues of war, poverty, and racism and with economic stability at home and economic relationships abroad. The effort to cope with air pollution as if it were a separate issue is one of the reasons for the failure of control. An approach to air pollution as part of the whole environmental crisis would make possible the fundamental changes that are the only hope for reducing the burden. Within the framework of this book, it is possible to focus on only one environmental issue; others can be referred to only as they impinge on air pollution, but, in real life, air pollution cannot be dealt with in isolation. It is less obvious, but equally true, that the environmental issues cannot be dealt with in isolation from the other social issues. The environmental destruction of war dwarfs all assaults on the environment from peaceful agricultural and industrial pursuits; we cannot expect to create healthy cities without eradicating the present racist housing, employment, and educational patterns.

A New Basis for Determining Goals and Standards

In the SCEP (Study of Critical Environmental Problems) study of global pollution, the Work Group on Ecological Effects emphasized that:

> . . . before the end of the century we must accomplish basic changes in our relations with ourselves and with nature. If this is to be done, we must begin now. A change system with a time lag of ten years can be disastrously ineffectual in a growth system that doubles in less than fifteen years.[31]

Yet, in general, the recommendations in the report—most of which were for additional monitoring and research—were not in accord with this sense of urgency. Change was called for in regard to a pollutant only if:

(1) the fact that it is a key pollutant with harmful global effects has been established with a sufficient approximation of certainty or degree of probability to warrant remedial action, or
(2) informed scientific and professional opinion, or public and political opinion, or both, view it with sufficient apprehension or concern to warrant appropriate measures.[32]

Although these criteria do not require absolute certainty before remedial action is taken and do allow some scope for scientific and public concern even without proof of harm, they also permit a continuation of the past do-nothing approach to global pollution until additional facts are discovered. If the words "human health effects" were substituted for "global effects," point (1) above would also describe very well the basis of our present city-oriented control program. We take our economy and our technology, with its resultant pollution, for granted and do not change them unless the harmful effects of the pollution on human health are established "with a sufficient approximation of certainty or degree of probability" to warrant action. This might be considered a conservative approach, and so it is—conservative of the economic system. A more fundamentally conservative approach would be one that is conservative of the human species and the environment. "Instead of dealing with environmental problems as a subsystem of the economic system, we must regard the economic system as a subsystem of Man's relationship with Nature," French economist Bertrand de Jouvenel told an international symposium on the problems of the environment in May 1971[33].

If we take our natural environment as the starting point—accepting the world as it can be described by biology and meteorology, rather than by economics and technology—the wording of the SCEP criteria would need to be altered very little, but we would have a different definition of "change": Change is the *addition* to the natural world of any pollutant from man's activities, instead of any *removal* of such a pollutant. We could then say that change is permitted only if

> the fact that it will not interfere with environmental processes or cause harm to people has been established with a sufficient approximation of certainty or degree of probability to warrant allowing its release.

This would be a simple, but profound, difference in our way of looking at air pollution. It would deny that there is any right to use the atmosphere for waste disposal. It would place the burden of proof on the polluter instead of on the pollution-control mechanism. It would consider a pollutant guilty until proved innocent, rather than vice versa.

We have taken a small step in this direction with the requirement that government agencies prepare "environmental impact statements" describing the impact on environmental quality of a proposed project and the alternatives considered, thus opening the way for environmental considerations to become part of the decision-making process.

We also have a precedent in the Pure Food and Drug Act, which requires proof of harmlessness from a manufacturer before permitting a drug or food additive to be put on the market. This system has not prevented numerous problems with food and drugs, suggesting that turning the tables of proof would be only the beginning and not a solution in itself. Uncertainties would still have to be dealt with, although they would be approached from a new angle, emphasizing prudence, rather than recklessness, in interfering with the natural environment and requiring changes —some difficult and expensive ones—in human activities on the basis of incomplete evidence of harm, rather than waiting for hard data in the shape of measurable increases in illness and death. There would still be room for judgment and for differences of interpretation. If we attempted to apply such criteria to automotive pollution today, for example, we would find spokesmen for the manufacturers maintaining that the harmlessness of carbon monoxide in present emissions has been established with a *sufficient approximation of certainty* to warrant allowing it to be continued [34], while many medical scientists and control officials would disagree. A system based on this principle, like the present one, would require adequate staffs and adequate funding in regulatory agencies and a public which is alert to the influence of special interests and informed on technical questions.

Ultimately, whose judgment prevails depends upon social organization, the power wielded by special interests, the effectiveness of the channels for public participation in decision making. This further emphasizes that neither air pollution in particular nor environmental problems in general can be solved separately from the basic questions of political and economic power.

The principle seems to imply a goal of *zero pollution,* long considered the ultimate in utopian unreasonableness since all human activity produces some pollution. It is quite true that a completely unpolluted atmosphere, like any other form of perfection, is impossible to attain. The question is whether it is reasonable to seek to approach it as closely as is humanly possible or, in fact, whether we have any other choice in the long run.

Scientific Advance and Technological Change

The depth of our ignorance and the extent of our uncertainty on such questions as how the earth-atmosphere system works, how we are affecting it by using the atmosphere for waste disposal, and, in turn, how this waste in the air is affecting the lives and health of people have been discussed previously. The urgent, but patient, pursuit of the knowledge we need so badly is the task of science. Building on this knowledge by developing ways of controlling or eliminating emissions, of recycling the wastes that now are dissipated in the air, of altering the extraction, transportation, and manufacturing processes to lessen their undesirable impact on the atmosphere and related biological systems—these are the tasks of technology. But science and technology are not separate enterprises; they are part of our whole social enterprise.

Science has not one frontier, but many. Where any individual scientist or group of scientists seek to push out into the unknown depends upon many things. Our society prides itself on the freedom of scientific inquiry it provides, but it is not the individual scientist's interests and values alone that determine where and how he works, but also those of the society that poses the problems and provides the financial support. A scientist may follow a well-beaten track whose extension into the wilderness has already been blazed by advance scouts, some of whom are his teachers or associates; he may join a well-financed wagon train which is recruiting for the next expedition; he may be attracted to a newly discovered trail by the excitement such a discovery always engenders; he may be stimulated by what he sees as an urgent social need to break through at a particular point; he may be deflected from a preferred goal for lack of financial support or because he can't go it alone and others are uninterested in moving in the same direction. Few scientific inquiries these days can be pursued by a single individual and, when it comes to the environment, the inquiry must be not only collective, but interdisciplinary. In short, the scientist is part of a whole scientific endeavor and that endeavor is embedded in the whole of society—ethically, economically, politically, organizationally.

This is even more obvious when it comes to technology for, much more directly than science, technology gets its marching orders from government or industry and is geared to the accomplishment of a particular task.

It is the accomplishments of particular tasks without regard to their side effects that have created the environmental problems we now have: Today's power plants, automobiles, and manufacturing processes are highly successful in terms of the jobs technology set out to do; unfortunately, the specifications did not include consideration of air pollution or other environmental problems. When technology is given a task related to clearing the air, such as removing particles or sulfur oxides from stack effluent, again the specifications are narrow. What happens to the pollution after its removal is usually not part of the problem. Will a wet scrubber dump the particles it removes from the stack into the water, thus contributing to water pollution? Will the sulfur be reused or will it be removed from the air only to be discarded elsewhere? Utilities or manufacturing industries that are required to regulate their effluents are interested in meeting the regulations in the most economically feasible manner, and current emission standards give them that freedom.

As we have already noted, the uniform national emission standards for new installations in a number of industries look toward a time when all plants in operation will have the best available, economically feasible control technology. It might be asked why this—or an extension of it—is not good enough. How can we control pollution any better than the best available technology permits? The answer lies in three other questions: What are the criteria for the "best available" control? For economically feasible technology? Under what circumstances is new—and better—technology developed?

There was *no* technology for the removal of sulfur dioxide from power plant stacks until sulfur dioxide standards began to be set. There are now several removal processes, but they are not yet "commercially available" because they have operated successfully only on small pilot plants.

Moreover, control technology must pass the test of economic, as well as technical, feasibility. This is an indefinite criterion, but one which generally means a process which will not interfere seriously with present production processes and market relationships. It does not refer to the most economical process in the long run from the social point of view, but to the most economical process in the short run from the point of view of the specific private industry affected.

There are two different kinds of sulfur oxide removal processes now on the road between development and commercial availability—those which recover the sulfur and those which throw it away. If they clean stack effluent equally well—or at least are both able to meet the emission standards—it would be left to the company owning the stacks to select the process it prefers. This might well be one of the throwaway processes because they are lower in capital cost and therefore more economically feasible for existing boilers[35].

But considerations of which process is more economically feasible in the long run, and from a social and biological point of view, would indicate the selection of a process which recycles the sulfur. Athelstan Spilhaus pointed this out in a pertinent passage on the economics of collecting pollutants at the source:

> Regardless of what any economist tells me, I'm convinced by the second law of thermodynamics that it must be cheaper to collect something at the source than to scrape it off the buildings, wash it out of the clothes, and so forth.
>
> The second law of thermodynamics tells us—among other things—that, as a product moves through a series of changes, it loses some energy in each change. In order to reconstitute the product in a more complex, more useful form, new energy must be introduced—as heat, as man-hours of work, or in some other form. Even if what we want to do is simply regain parts of the original product that have been dispersed, rather than actually reconstitute it, this is true. The earlier in the series of changes the new energy is introduced, the less energy will have been lost and the less new energy is required. Therefore, it must be cheaper in energy (and energy costs money) to collect waste at the source before it has undergone a long series of changes and dispersals. This is a kind of complete rethinking in economics.[36]

It is a kind of rethinking, however, that has not yet penetrated either our economy or our technology.

There was no progress in automobile emission control until California made it clear that uncontrolled engines would no longer be permitted to travel California roads. National emission standards followed, demanding better control than any presently available technology can provide. Stimu-

Pollution is a resource misplaced. Whether we want to capture the resource or avoid the unwanted consequences of the pollution, or both, it is cheaper in energy and money to prevent the resource from becoming pollution rather than to disperse it into the air and then have to scrape it off buildings, wash it out of clothes, and so on. This photograph shows 34 years of dirt being removed from Cincinnati City Hall in 1963.

lated by these regulations and observing a cloud on the horizon in the shape of government-sponsored research on alternate propulsion systems to the internal combustion engine, car manufacturers in Detroit are finally taking emission control seriously. However, engineering considerations suggest that better technology was at hand all the time—"better," that is, from the point of view of reducing pollution, but requiring a sacrifice in speed. A car with a low-speed, low-compression engine, running on low-octane fuel and therefore requiring no lead additives in its gasoline, adjusted to operate best at about 40 miles per hour (although able to achieve higher speeds), could easily have been produced by adapting pre-war models and would probably have been much less polluting [37]. This route to reduced automotive pollution was never tried, presumably because bigger, more powerful cars were assumed by the manufacturers to represent "better" technology from the point of view of sales and profits.

It is part of the conventional wisdom about our economic system that the world will beat a path to the door of the man who invents a better mousetrap. A Department of Commerce report strongly suggested that, compared to the internal combustion engine, the Rankine external combustion engine, a steam car, is a "better mousetrap"—a satisfactory alternative in terms of performance and far superior in terms of emissions [38]. Then why isn't the Rankine engine a competitive threat to the internal combustion engine? The same report made it clear that the reasons were economic. The established automobile industry has too much invested in the internal combustion engine to be interested in developing an alternative, and the costs for an independent manufacturer to enter the market are almost prohibitive. In addition to the enormous initial investment, there would be the difficulties of establishing dealerships and servicing facilities. The petroleum industry also presents an obstacle, since its refineries would have to be revamped to produce low-grade kerosene instead of the present high-octane fuels. (However, note that in the long run this would be a saving.)

Even the present air pollution control efforts have been able to stimulate the development of technological improvements, but, in most cases, within the limits of tack-on devices which seek to abate the release of pollution *after* it has been produced. More basic technological change which seeks to alter a process so that less pollution is produced is much slower in developing and, once developed, may not be put into use. The Department of Commerce report on the Rankine engine makes it quite clear that we no longer have a free market in the automobile industry where it would be possible for one company to challenge the established giants by offering the public a low-pollution vehicle.

Industry invests tremendous sums for research and development when increased sales and profits or new processes which will cut labor costs are in view. Such incentives have been responsible for the development of some of the air pollution monitoring equipment and the tack-on air pollution control devices now or soon to be in use. But thus far these incentives have produced little in the way of process changes which would cut down on the production of waste or closed systems which would completely prevent the release of waste to the environment. One of the most hopeful developments in the electric utility field is a process for generating electricity from coal which promises to be both more efficient and less polluting. In introducing a discussion of this "fluidized-bed" technique as a possible replacement for the standard "pulverized fuel" (PF) technique now in use, the leading U.S. proponent of the former, Arthur M. Squires, implied that it had not been developed and put into use on a large scale because the technological community does not yet really accept the idea that air pollution controls are imperative.

"After the acceptance," he says, "everyone may be pleasantly surprised at the innovations mothered by necessity."[39] Clean power, Squires believes, may become available from coal at a profit. What is lacking is a social mechanism to mother the recognition of necessity which will, in turn, mother the innovations.

The obstacles to a changed process for coal are similar to those already noted to a changed engine for the family car: There is a mature, well-established technology to which the present utilities are committed. "It is much harder to overturn an established technique, particularly when equipment has grown to large sizes, than it is to introduce an entirely new technology," Squires notes[40]. Competition to put the cleanest and most efficient generating process on the market has been restricted by the common ownership of the various fuel sources. For example, the Continental Oil Company and Gulf Oil Corporation have acquired large coal companies and are in the nuclear energy market as well. Six other oil companies also own large coal reserves[41]. One of them, Kerr-McGee, also has big uranium mining interests.

Government is sponsoring some research in coal engineering, but is much more heavily committed to research on fast-breeder nuclear reactors. In 1970, the federal government spent about $367 million for research and development in the field of energy, but about 84 percent of that was for nuclear energy. The electric power industry devotes less than a quarter of a percent of operating revenue to research and development[42].

More Efficient Use of Energy and Other Resources

In every energy transformation some heat is lost; whenever matter is burned, some waste products are released. A serious effort to reduce air pollution—thermal, gaseous, and particulate—therefore requires investigation into the way we are using our energy resources with an eye to cutting down on both kinds of waste. The finite nature of both stored energy and stored materials calls for the same kind of inquiry.

What are the most efficient and least polluting methods for generating electricity? How can the waste heat of generating plants be utilized? What industries are the greatest consumers of energy? Can their products be replaced by others using less energy? How can we transport people and goods more efficiently, using less energy and producing less pollution? There are no simple answers to any of these questions; much could be written about what is now known about the answers, while simply asking the questions in this context suggests directions for further study and the kind of social decisions that are still to come.

Research and development is well funded only for nuclear power, but it is far from clear that all our eggs belong in the nuclear basket or that nuclear research should be so heavily committed to fast-breeder reactors at the expense of fusion power[43].

There are other possibilities for the use of fossil fuel besides the fluidized-bed technique already mentioned. Furthermore, the efficiency of steam-driven turbines could be improved, whether they obtain their power from fossil or nuclear fuel, by topping the steam cycle with a gas turbine or magnetohydrodynamic (MHD) generator. Geothermal, solar, wind, and tidal energy are sources which have been slighted because they are applicable for only certain uses or in particular places, but they might contribute in an important way to total energy needs and produce less pollution than either fossil or nuclear fuel. They therefore deserve more research support than they have received thus far. Solid waste can be used directly as fuel (as has been done in Paris and other European cities for years) or can be processed to produce fuel for generating electricity[44].

Because generating electricity with coal and oil creates more pollution, gas will be widely used by electric utilities in the next few years. At the same time, coal- and oil-generated electricity is being used increasingly for space heating and cooling, although gas would do equally well. Is it really efficient to transform gas into electricity, losing some heat in the process, rather than allocating the cleanest natural fuel directly to space heating and cooling? One possibility for the use of gas in the future is to have fuel cells operating on natural gas built directly into dwelling units to generate the household's electricity on the spot. Operating at a higher efficiency and lower temperature, with no need for transmission lines and only carbon dioxide and water vapor as waste products, the fuel cell has much to recommend it[45] and deserves the attention needed to work out its technical and economic problems as rapidly as possible.

Waste heat is most often discussed as a water pollution problem, but, whether it is discharged from a power plant or other industry directly into the air or indirectly via a stream or lake, sooner or later it becomes part of the waste heat that must be rejected into the atmosphere, producing the growing effects on global heat balance and climatic change discussed earlier. It is therefore an air pollution problem in this sense, while, if we can find uses for some of the waste heat, we will be improving the efficiency of our energy use as well as reducing the amount of other forms of pollution produced.

Recently, the Oak Ridge National Laboratory of the Atomic Energy Commission studied such possibilities in an imaginary new city of 389,000, with a power plant built to serve it, and found that "we could prevent the

Cooling towers like these can prevent the destruction of aquatic life in rivers and lakes that occurs when warm water is discharged from a power plant. But they may also affect local weather. Whether waste heat is discharged directly into the air or into water first, sooner or later it is rejected into the atmosphere where it may affect regional and even global heat balance.

burning of over four million barrels of oil each year with its accompanying air pollution, and reduce thermal effluent at the power station by 40 percent." [46] (The overall reduction in waste heat was the amount that would have been generated by the four million barrels of oil, for the 40 percent reduction at the power plant is significant only as it relates to the *concentration* of waste heat. After being used for space heating, this same waste heat is rejected to the atmosphere from a number of more diffuse sources.)

Waste heat from electrical generating plants is already used for space heating, for example, in Tapiola Garden City in Finland, with a population of 20,000 [47]. Other uses for waste heat suggested by the Oak Ridge study were industrial heating, renovation of sewage water by distillation, and greenhouse heating for food production.

An industry-by-industry examination of energy consumption and environmental impact from the point of extraction of raw materials to the use of the finished product and its disposition as waste would give us some needed insight into the way we are exploiting our environment both for its resources and by using it as a sink for waste—sometimes for end products that may contribute disproportionately little to our well-being.

The primary metals industry is by far the largest industrial consumer of electricity. Much of this, in turn, is accounted for by aluminum production which consumes about ten percent of all industrial power use. Manufacture of aluminum is expected to continue to increase, especially for automobiles, trucks, buses, and containers. All of these now wind up as serious solid waste problems.

Or take the paper industry. Its environmental impact begins with its timbercutting practices and the cost of timber replacement, and extends through a highly-polluting manufacturing process which uses large amounts of electric power to the final burning of thousands of tons of paper every day throughout the country. One question that is now being asked quite insistently is: "How much of this paper can be recycled?" Another question that needs to be asked is, "How much of it do we really need?" New technology is cutting down on the pollution at the manufacturing stage, but this is being offset by increases in production.

We utilize a tremendous amount of energy for transportation in the least efficient way conceivable: 360 to more than 400 horsepower four-, six-, or eight-passenger cars, often carrying a single individual. These cars are

The environmental impact of the paper industry begins with its timber-cutting practices and ends with the burning of thousands of tons of paper daily, much of it newsprint. Shown below is the newsprint consumption from 1950–1968, broken down for the amount used in news, advertising, and non-newspaper uses. Recycling of paper is only a partial solution. Considerations of the social usefulness of the product do not affect the decisions by manufacturers, suppliers, and customers which determine how much newsprint is produced, nor is the industry's environmental impact considered, nor are the two factors balanced against one another.

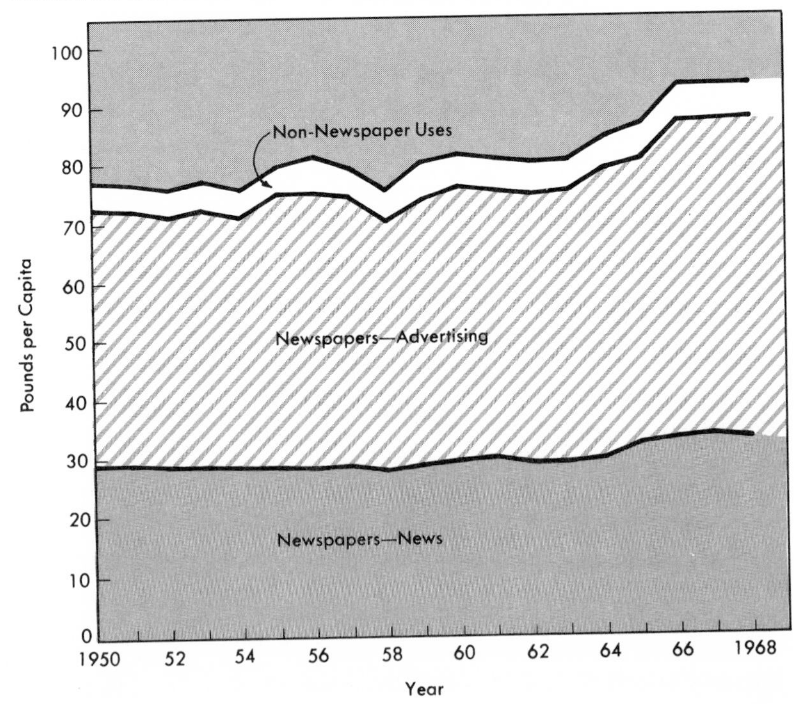

not only inefficient and polluting, they are also killing thousands of people every year in traffic accidents. New, less polluting technology for the private car is therefore only a small part of the answer. When many people are moving in the same direction at the same time, mass transit could do the job far more efficiently, safely, and with a small fraction of the pollution. We are also inefficient in moving freight. Per ton mile of freight haulage, trucks burn nearly six times as much fuel as railroads and emit about six times as much pollution [48].

Although power use per capita is much higher in the developed countries with a high Gross National Product (GNP), the correlation between the two is not as close as is usually assumed. In an interesting assessment of energy and materials utilization, two University of California scholars showed that while the United Kingdom, Belgium, Australia, Germany, Denmark, Norway, France, and New Zealand had GNPs within ten percent of each other, energy consumption per capita for industry, commerce, and transportation varied betweed 45 million BTU in New Zealand and 110 million in the United Kingdom [49].

While the United States could not maintain its present standard of living by emulating the agricultural economy of New Zealand, it could maintain it at a very much lower per capita energy use, according to this analysis, by such measures as more efficient energy generation, shifts from more to less energy-intensive materials, carrying more freight by rail and less by truck, and reusing and recycling various products.

Revising Our Outlook on the Growth Rate

Nowhere is the contradiction between an ecologically balanced civilization and our present unbalanced civilization so sharp as on the question of growth. Expansion of the economy is the national goal in socialist as well as capitalist countries, in underdeveloped as well as developed countries. Estimates of increases in U.S. air pollution cited earlier in this chapter, as well as those for global waste heat and atmospheric particulates given in earlier chapters, have all been predicated on continued economic expansion. Yet economic growth is clearly not possible indefinitely on a planet with finite space and resources, and those inexorably rising pollution curves could level off and even drop if there were some way to slow or halt economic growth. The question is, can this be done without ruining the economic system and destroying the livelihood of the people who depend upon it?

Of the many facets of growth, population has received the greatest attention. Yet even if we could stabilize the population of this country immediately at its present level, we would not have touched the heart of the problem. Increasing rates of power use are an important aspect of economic growth in our society, and they are only 20 percent dependent upon population increase. Per capita increase in power use is responsible for the other 80 percent [50]. This does not mean, of course, that individual families are increasing their direct consumption of power at that rate. Industrial use accounted for 41 percent of total electric power production in 1970. Residential and commercial uses divided the remainder about equally after allowing for ten percent loss, much of it in transmission[51]. Each increment of economic growth does not place the same stress on resources or create the same amount of pollution[52]. Nevertheless, under the best of circumstances there is some upper limit to growth. An ecologically balanced civilization means, eventually, no further growth in the economy and no further growth in the population.

Proposals to halt economic growth generally meet with two objections. First, economists claim that dealing with pollution problems will require more, not less, energy and production and that growth is essential to the operation of our economic system. Electrostatic precipitators (the collectors of particulates from power-plant stacks), for example, are enormous, expensive contraptions almost as big as the plants themselves. Manufacturing many such units obviously requires both materials and energy, and more energy is required to operate them. But one of the reasons for the high energy requirement is our insistence on regarding control as something which takes place at the *end* of the industrial process, rather than at a much earlier point or, whenever possible, at the very beginning. Furthermore, the energy required by pollution control has been exaggerated, as a recent analysis of the energy needs of waste-water treatment shows[53].

The second economic objection is more fundamental, for essentially what it says is that our economic system is incapable of stability: It must grow or recede until the recession makes possible renewed growth. This analysis appears to be borne out by the boom-and-bust history of the system in the United States. Although the years since World War II have seen more consistent economic growth and briefer, less severe recessions than any previous period, these years have also seen a phenomenal rise in pollution and a high proportion of productive resources devoted to nonproductive war material. From an environmental, as well as a human, point of view, the latter is worse than nonproductive; it is both wasteful and destructive. The need for growth drives the economy to other forms of waste. Production becomes important for its own sake; the use to which the product is put is important only to the extent that this affects purchase. Advertising is relied upon to stimulate ever-increasing purchases and short-lived products to make a new round of purchasing necessary. Economists now claim that depressions are no longer inevitable, but this claim is based on the view that we now have ways to keep the economy growing—not that we have learned how to stabilize it at a given level.

Certainly there are economists and environmentalists who believe that this economic system has the flexibility to adapt to environmental needs, although few have faced the basic nature of the changes required. The solution to our environmental problems will not be easy, will not come automatically, and will require important economic readjustments.

Making the Economy More Responsive to Environmental Concerns

"Environmental problems stem largely from [the] fundamental failure of the economic system to take account of environmental costs," so says the President's Council on Environmental Quality in its 1971 report[54].

The general direction of change suggested by the Council is to find ways to include in the prices of the products produced the social costs of the wastes discharged:

> When the full production costs are included in the prices of final products, the market allocates resources efficiently. If, however, all costs are not included—for example, the costs to society of environmental degradation—then the resulting prices of the products are too low. When products are underpriced, consumption of them is higher than it would be if all costs were included. Consequently, compared with other products, too many resources are devoted to their production. To the extent that the costs of preventing undesirable environmental impacts are not reflected in the price of goods and services, the market fails to allocate resources efficiently, and too much waste is produced.[54]

We have already seen that this confidence in the operation of the market is somewhat questionable. But even if it were justifiable, how is the pricing system to be adjusted to include social costs? One way would be to require controls (on automobiles, for example) which increase the price of the product. Another would be the proposal for an "effluent fee" or "pollution charge"; for example, the President has proposed a charge on sulfur oxide emissions to make pollution more costly for utilities and manufacturers than abatement.

The rationale for this approach is that the subsequent improvement of environmental quality would be shared by everybody, while the cost would be borne by those paying the higher "real" costs for some goods and services. But would there not be another social cost to such a solution, and one which would weigh most heavily on those least able to pay? It has already been estimated by a government air pollution control official that the cost of new cars with control devices, plus the added cost of maintaining those controls, will price some low-income customers out of the market.[55] In cities built around the automobile, how are these people to get to work? To shopping centers? To places outside the cities where they can enjoy the improved environment? Unless higher-priced cars are supplemented by mass transit, the poor will bear a disproportionate share of the economic cost of cleaner air, just as they now bear a disproportionate share of the physical cost of dirty air (see pages 139–40).

Enterprises which find the cost of pollution control too high frequently threaten to go out of business. Although this sometimes proves to be simply an effort to forestall controls, a number of small enterprises have actually closed. For example, the open burning of boxcars in a salvage operation in Illinois was outlawed because of the air pollution it produced. When the

Inefficient, polluting, lethal—but individually convenient—the private passenger car has become part of the American way of life and death. Less polluting technology for the private car is only part of the answer. More efficient, less dangerous methods of mass transportation, supplemented by the private car in a less central role, could cut down on the amount of energy used per passenger mile, on the amount of pollution emitted, and on traffic accidents and deaths.

state finally insisted on pollution control after the company had stalled for six years, the salvage operation shut down and 50 people lost their jobs. Unions have begun to call for special means to cushion environmental unemployment.

One environmentalist, writing in *Science,* proposes to carry the fee or tax system even farther in order to deal with urban sprawl and other aspects of growth:

> . . . to accelerate the discovery and exploitation of mineral resources, we now give generous depletion allowances. However, to encourage more efficient use of such resources, we may need to institute resource depletion taxes. We might also need a space depletion tax to encourage effective use of land to discourage our urban sprawl.[56]

However, a space depletion tax would exert only a negative control in an area where positive planning and collective social action seem necessary and where manipulation of market relationships is inappropriate because much more than buying and selling is involved. Municipal, regional, and national planning, on a scale to which we are quite unaccustomed, are basic to the proper design of transit systems, to efficient and environmentally sound land use, and to a rational energy policy. Mechanisms for making such plans and carrying them out are missing in our private-enterprise economy and our present political system.

It is also difficult to see how effluent taxes and fees could cope with the many different air pollutants, some of which are much more serious in combination with other pollutants than they are alone. The control mechanism would tend to become excessively complex, unwieldy, and difficult to administer.

Whatever path is taken to stimulate controls or regulate emissions, it appears inevitable that in the period of transition the cost of production will rise as industry begins to pay for formerly free waste disposal. Overall productivity will drop as more man-hours are devoted to pollution control, a previously missing aspect of production which adds nothing to total output. This will tend to lower the rate of profit and lower profits tend to depress wages or raise prices, or both.

Eventually, the true economy of reuse rather than waste will pay off. It is an unavoidable fact, however, that an economic system built to exploit seemingly boundless resources in air, water, land, and under the earth is not going to adapt easily to a limited resource base, production that is relatively pollution free, and stability rather than expansion.

Economist John Kenneth Galbraith believes that "environmental disharmony" is the consequence of the disparate goals of what he calls the "technostructure" (business and technology) and the public. He advocates discarding "existing economics" and developing a system in which the state would act as arbiter between the technostructure and the public. The state, says Galbraith, should control prices, protect the environment, and support housing, urban transport, municipal services, and culture [57].

Galbraith has challenged both the desirability of growth and the concept that it is secondary to profit seeking. The interest of business people, he says, "is most strongly served not by earning, but by growth" because growth brings them bigger salaries, more perquisites, and greater security, influence, and power. Businessmen manipulate the economy to produce more and more of whatever is most convenient, regardless of consumer or citizen interests[57].

Galbraith proposes a system which would represent a major change from private enterprise. It would require that those now having the power to manipulate the economy voluntarily relinquish some of that power to the state, or that the state, on which they now exert a large influence, take power away from them.

Since the ecological facts of life are the same in socialist and capitalist countries, we may look forward to some interesting competition as the two systems strive to solve their environmental problems. At the 1971 Prague symposium, spokesmen for both systems claimed a greater flexibility and capability for coping with pollution[58].

Socialist theoreticians are quick to point out the environmental problems inherent in capitalism. Before the turn of the century, Frederick Engels, who recognized that every technological advance exacted a price from nature, claimed that a capitalist system is able to concern itself only with immediate profit, not with "what becomes of the commodity afterwards" nor with the effects on nature of the production process [59].

A socialist system may have the potential for a technology designed to take the cost to nature into consideration, but in the Soviet Union other shorter-term considerations have dominated successive national plans: First it was laying the base for an industrial society, then repelling the Nazi invasion, and, finally, rebuilding a war-devastated economy. Since the war, a considerable portion of Soviet industrial resources has been devoted to the production of war materials and a sizable fraction of scientific and technological effort to military and space technology. The result of 50 years of industrial development under socialism has therefore produced some of the same problems that plague industrially-developed capitalist countries.

While dispersion is still being advocated as a solution to industrial air pollution in both the U.S.S.R.[60] and the United States, there are differences as well as similarities in both problems and solutions. The Russians have gone ahead with the SST in spite of its environmental side effects. On the other hand, photochemical smog is unlikely to become as serious a problem in Soviet cities as it is in American cities. Although production and use of the private automobile is increasing in the Soviet Union, Soviet cities are planned around a minimum, rather than a maximum, use of the automobile. Soviet urban design is based on neighborhoods or "microdistricts" in which everything ordinarily needed for family life is within a five- or ten-minute walk of every inhabitant, while public transportation from the neighborhoods to the central cities is rapid and inexpensive[61].

Both the U.S.S.R. and the United States have had enormous natural resource bases on which to build and are geared to indefinite expansion. Like the capitalist economy, the Soviet economy will face its major challenge in changing its concept of unending growth.

Even the most modest economic changes, such as the imposition of an effluent fee for sulfur in the air, must be politically initiated. The strength and the environmental goals of various conflicting political forces will determine the direction and extent of the economic changes we will make in the coming years and, therefore, whether or not we will succeed in reducing the air pollution burden in this country and its contribution to global pollution. These changes cannot be brought about by experts in economics and political science, any more than they can be brought about by experts in ecology or air pollution technology. Yet the services of experts

The new subway in Baku, capital of Azerbaijan in the U.S.S.R. Baku is one of five Soviet cities with subway systems. Expansion of these subways and the construction of new ones in Kharkov, Tashkent, Kuibyshev, and Gorki are planned or are under way. In cities of more than a million population, Soviet planners consider that subways should carry the main transportation load for reasons of speed and safety, because of the limited traffic capacity of urban streets, and because the subway helps to reduce fouling of the air and excessive street noise.

in all these fields are needed to provide a sound information basis for political action.

The need to cross the old disciplinary lines in research has already been discussed, but this alone is far from enough. Fruitful dialogue across the boundaries of the natural and social sciences is important, too. Beyond that, communication between natural and social scientists and ever-wider sections of the public is vital. The narrow professionalism and self-seeking careerism that inevitably develops in a society based on the exploitation of nature for the aggrandizement of individuals are totally inappropriate to a society that must learn how to adjust to nature and preserve it for all who live in it and will live in it far into the future. This is a task that requires a broad public understanding of man's relationship to nature, of the roots of our environmental crisis, of the alternatives for coping with it. No expert can—or need try to—tell the public what course to take. But any man or woman who embarks on a career in natural or social science or in technology and fails to make his skill and his knowledge available to the public will be irrelevant not only to one of the most challenging social problems of the time, but to the survival of the human species.

More than any other environmental problem, air pollution is obviously global. Like disciplinary boundaries, national boundaries too must be crossed if we are going to learn how to live together in our one world with its undivided atmosphere.

When the history of the second half of the twentieth century is written, we may all be judged most severely by whether we bequeathed to our descendants the almost insuperable task of restoring a planet ravaged by manmade disasters or whether we recognized the danger to the one indivisible atmosphere and earth and acted in time to pass on to the twenty-first century a world in which life was still livable.

[1] Paul W. Spaite and Robert P. Hangebruck, "Pollution from Combustion of Fossil Fuels," *Air Pollution—1970, Hearings before the Subcommittee on Air and Water Pollution of the Committee on Public Works, U.S. Senate,* Part 1 (Washington, D.C.: U.S. Government Printing Office, 1970), p. 173. Emphasis added.

[2] Ibid., p. 175.

[3] Vincent J. Schaefer, personal communication, December 3, 1971.

[4] A. Eugene Vandegrift et al., "Particulate Air Pollution in the United States," *Journal of the Air Pollution Control Association,* June 1971, 21: Fig. 2, p. 323.

[5] Vincent J. Schaefer, "Some Effects of Air Pollution on Our Environment," *BioScience,* October 1969, Vol. 19, No. 10.

[6] Sheldon Novick, "And for Our Next Number . . .", *Environment,* June 1971, Vol. 13, No. 5, 29.

[7] *Chemical & Engineering News,* March 8, 1971, 14–15.

[8] *Motor Vehicles, Air Pollution, and Health,* Report of the Surgeon General to the U.S. Congress, House Document 489 (87th Congress), U.S. Department of Health, Education and Welfare, Public Health Service, Division of Air Pollution, Washington, D.C., 1962.

[9] *Federal Register,* April 7, 1971, 36:67, for projections to 1985. APCO provided the extension to 1990 in a personal communication, July 30, 1971.

[10] Staff report, "Report Card," *Environment,* September 1970, Vol. 12, No. 7, 24–25.

[11] D. S. Barth et al., "Federal Motor Vehicle Emission Goals for CO, HC, and NO_x Based on Desired Air Quality Levels," *Air Pollution—1970,* Part 5, p. 1641 (see [1]).

[12] J. R. Coleman and R. Liberace, "Nuclear Power Production and Estimated Krypton 85 Levels," *Radiological Health Data and Reports,* November 1966, 7:615–21.

[13] "Cooling It in Minnesota," *Environment,* March 1969, Vol. 11, No. 2, 21–25.

[14] Robert Gillette in "News and Comment," *Science,* June 18, 1971, 173:1215–16.

[15] John J. Russell and Paul B. Harn, "Public Health Aspects of Iodine 129 from the Nuclear Power Industry," *Radiological Health Data and Reports,* April 1971, 12:189–94.

[16] Eric J. Cassell, "The Health Effects of Air Pollution and Their Implications for Control," *Law and Contemporary Problems,* School of Law, Duke University, Durham, North Carolina, spring 1968, Vol. XXXIII, No. 2, 197.

[17] Clean Air Act as amended in 1970, Section 112(a)(1).

[18] Victor H. Sussman, Director of Pennsylvania's Air Pollution Bureau, speech reported in the *St. Louis-Post Dispatch,* January 14, 1970.

[19] Robert E. Kohn, "Linear Programming Model for Air Pollution Control, A Pilot Study of the St. Louis Airshed," *Journal of the Air Pollution Control Association,* February 1970, 20:82.

[20] Albert E. Smith and Bernard Bloom, "Air Quality Standards," Letter to *Science,* May 12, 1972, 176:581.

[21] George B. Morgan et al., "Air Pollution Surveillance Systems," *Science,* 170:293.

[22] Environmental Protection Agency press release, April 30, 1971.

[23] George E. Hagerik, *Decision Making in Air Pollution Control* (New York: Frederick A. Praeger, Inc., 1970).

[24] *Air Pollution—1970,* Part 1, p. 252 (see [1]).

[25] Stanford Workshop on Air Pollution, *Air Pollution in the San Francisco Bay Area,* Stanford, California, summer 1970. This report gives a fascinating description of how industry and the control agency interact, excluding the public until compromises have been made and technical details worked out.

[26] Henry Steck, "Why Does Industry Always Get What It Wants?" *Environmental Action,* July 24, 1971, 11–14. Dr. Steck is professor of political science at the State University of New York. The article is based on his testimony before the Senate Subcommittee on Intergovernmental Relations.

[27] John Esposito (ed.), *Task Force Report on Air Pollution,* Center for the Study of Responsive Law, Washington, D.C., 1970.

[28] Roy Craig, "Cloud on the Desert," *Environment,* July-August 1971, Vol. 13, No. 6, 32.

[29] Terri Aaronson, "World Priorities," *Environment,* July-August 1972, Vol. 14, No. 6, 4.

[30] Barry Commoner, *The Closing Circle: Nature, Man, and Technology* (New York: Alfred A. Knopf, 1971), p. 12.

[31] Study of Critical Environmental Problems (SCEP), *Man's Impact on the Global Environment* (Cambridge: The MIT Press, 1970), p. 126.

[32] Ibid, p. 244.

[33] David N. Leff, "A Meeting in Prague," *Environment,* November 1971, Vol. 13, No. 9, 31.

[34] David Bird, "Car Makers and Ecologists Are in Conflict," *New York Times,* May 23, 1971.

[35] *Air Pollution—1970,* Part 1, p. 173 (see [1]).

[36] Athelstan Spilhaus, "Waste Management and Control," *Scientist and Citizen* (now *Environment*) November-December 1967, 9:9–10, 220.

[37] John Macinko, "The Tailpipe Problem," *Environment,* June 1970, Vol. 12, No. 5, 6–13.

[38] Terri Aaronson, "Tempest Over a Teapot," *Environment,* October 1969, Vol. 11, No. 8, 23.

[39] David A. Berkowitz and Arthur M. Squires (eds.) *Power Generation and Environmental Change* (Cambridge: The MIT Press, 1971), p. 176.

[40] Ibid., p. 177.

[41] Rush Loving, Jr., "How Kennecott Got Hooked with Catch-22," *Fortune,* September 1971, 100.

[42] Philip Boffey in "News and Comment," *Science,* June 26, 1970, 168:1554–59.

[43] Dean E. Abrahamson and Arthur Tamplin, *Comments on the Atomic Energy Commission's Draft Environmental Impact Statement for the Liquid Metal Fast Breeder Reactor Demonstration Plant,* Scientists' Institute for Public Information, New York, September 23, 1971.

[44] Hinrich L. Bohn, "A Clean New Gas," *Environment,* December 1971, Vol. 13, No. 10, 4–9.

[45] There are other possible fuels for fuel cells. For a discussion of the fuel cell,

its technical problems, and its possibilities, see Terri Aaronson, "The Black Box," *Environment*, December 1971, Vol. 13, No. 10, 10–18.

[46] Sam E. Beall, Jr., "Reducing the Environmental Impact of Population Growth by the Use of Waste Heat," paper presented at the annual meeting of the American Association for the Advancement of Science at the Symposium on Reducing the Environmental Impact of Population Growth, Chicago, Illinois, December 26, 1970.

[47] V. Santala, "How District Heating Serves Finnish City of 20,000," *Heating, Piping, and Air Conditioning*, September 1966, Vol. 38, No. 9, 129–35.

[48] Commoner, *The Closing Circle*, p. 171 (see [30]).

[49] A. B. Makhijani and A. J. Lichtenberg, "Energy and Well-Being," *Environment*, June 1972, Vol. 14, No. 5.

[50] Frederick E. Smith, "Power Generation and Human Ecology," in Berkowitz and Squires, *Power Generation and Environmental Change*, p. 7, (see [39]).

[51] Committee for Environmental Information, "The Space Available," *Environment*, March 1970, Vol. 12, No. 2, 2.

[52] This question is discussed at greater length by Barry Commoner in Chapter 9 of *The Closing Circle* (see [30]).

[53] Charles A. Washburn, "Clean Water and Power," *Environment*, September 1972, Vol. 14, No. 7, 41–43.

[54] Council on Environmental Quality, *Second Annual Report, August 1971* (Washington, D.C.: U.S. Government Printing Office, 1971), p. 102.

[55] *Environment Reporter: Current Developments*, July 16, 1971, Vol. 2, No. 11, 308, quoting Erik Stork, acting Director of the EPA's Bureau of Mobile Source Pollution Control.

[56] J. Alan Wagar, "Growth versus the Quality of Life," *Science*, June 5, 1970, 168:1184.

[57] From a report of a lecture given by John Kenneth Galbraith in Paris, *St. Louis Post-Dispatch*, February 19, 1971.

[58] Leff, "A Meeting in Prague," pp. 30, 32, 33 (see [33]).

[59] Frederick Engels, *Dialectics of Nature*, translated and edited by Clemens Dutt (New York: International Publishers Co., Inc., 1940), pp. 291–95.

[60] AP story from Moscow, *St. Louis Post-Dispatch*, August 31, 1971.

[61] A. Allen Bates (Office of Standards Policy, Department of Commerce), "Low Cost Housing in the Soviet Union," in *Industrialized Housing*. Materials compiled and prepared for the Subcommittee on Urban Affairs of the Joint Economic Committee, Congress of the United States (Washington, D.C.: U.S. Government Printing Office, 1969).

APPENDIX

Ambient Air Standards and Margins of Safety

Pollutant	Lowest Concentration Found to Produce Adverse Health Effects	Primary Standard	Margin of Safety
Sulfur Dioxide (Particulates also present; therefore the effect is not from sulfur dioxide alone, but from the combination.)	115 $\mu g/m^3$* (0.44 ppm*)[1] annual arithmetic mean particulates: 160 $\mu g/m^3$	80 $\mu g/m^3$ (0.03 ppm) annual arithmetic mean	30%
	300-1500 $\mu g/m^3$ (0.11-0.57 ppm)[2] 24-hour mean present for three to four days with moderate to high particulate levels	365 $\mu g/m^3$ (0.14 ppm) maximum 24-hour concentration not to be exceeded more than once a year	None-76%
Particulates (Sulfur dioxide present also; again it is the effect of the combination.)	80-100 $\mu g/m^3$ [3] annual geometric mean sulfation: 0.11 ppm	75 $\mu g/m^3$ annual geometric mean	6-25%
	300 $\mu g/m^3$ [4] 24-hour average sulfur dioxide: 0.19	260 $\mu g/m^3$ maximum 24-hour concentration not to be exceeded more than once a year	13%
Carbon Monoxide	12-17 mg/m^3* (10-15 ppm)[5]	10 mg/m^3 (9 ppm) maximum eight-hour concentration not to be exceeded more than once a year	17-41%
	58 mg/m^3 (50 ppm)[6] 90-minute exposure	40 mg/m^3 (35 ppm) maximum one-hour concentration not to be exceeded more than once a year	31%
Photochemical Oxidants	200 $\mu g/m^3$ (0.1 ppm)[7] maximum daily value	160 $\mu g/m^3$ (0.08 ppm) maximum one-hour concentration not to be ceeded more than once a year	20%
Nitrogen Dioxide	118-156 $\mu g/m^3$ (0.063-0.083 ppm)[8] mean 24-hour concentration over six-month period	100 $\mu g/m^3$ (0.055 ppm) annual arithmetic mean	15-36%

* $\mu g/m^3$ = micrograms per cubic meter
ppm = parts per million
mg/m^3 = milligrams per cubic meter

[1] A. J. Wicken and S. F. Buck, "Report on a Study of Environmental Factors Associated with Lung Cancer and Bronchitis Mortality in Areas of North East England" (London: Tobacco Research Council, Research Paper 8, 1964).

[2] L. J. Brasser, P. E. Joosting, and D. Von Zuilen, "Sulfur Dioxide—To What Level Is It Acceptable?" (Delft, The Netherlands: Research Institute for Public Health Engineering, Report G-300, July 1967).

[3] Warren Winkelstein, Jr. et al., "The Relationship of Air Pollution and Economic Status to Total Mortality and Selected Respiratory Mortality in Men," *Archives of Environmental Health,* 1967, 14:162–71.

[4] P. J. Lawther, "Climate, Air Pollution, and Chronic Bronchitis," *Proceedings of the Royal Society of Medicine,* 1958, 51:262–64.

[5] R. R. Beard and G. Wertheim, "Behavioral Impairment Associated with Small Doses of Carbon Monoxide," *American Journal of Public Health,* November 1967, 57:2012–22.

[6] Ibid.

[7] N. A. Renzetti and V. Gobran, *Studies of Eye Irritation Due to Los Angeles Smog, 1954–1956* (San Marino, California: Air Pollution Foundation, July 1957).

[8] Martin E. Pearlman et al., "Nitrogen Dioxide and Lower Respiratory Illness," *Pediatrics,* February 1971, 47:391–98.

Photo Credits and Copyright Acknowledgments

CHAPTER 1

Pp. 2–3, "Smog over St. Louis," photograph by Robert C. Holt, Jr., *St. Louis Post-Dispatch.*

P. 4, Episode 104 adapted from the Episode 104 map issued by the Environmental Science Services Administration.

P. 7, illustration by Robert Charles Smith for "Episode 104," *Environment,* January–February 1971.

P. 8, *Environment* photographs by Daniel T. Magidson.

Pp. 10–11, photograph courtesy of A. Devaney, Inc.

P. 13, photograph by Darwin Van-Campen, DPI.

P. 17, photograph by Robert C. Holt, Jr., *St. Louis Post-Dispatch.*

P. 19, photographs copyright © the *New York Times,* courtesy of NAPCA.

P. 25, *Environment* photographs by Robert Charles Smith.

CHAPTER 2

Pp. 30–31, "Monsoon in India," photograph by Brian Brake, Rapho-Guillumette.

P. 32, photograph courtesy of NASA.

P. 37, from "Air Pollution and Climatic Change" by J. M. Mitchell, Jr., presented at the 64th Annual Meeting, American Institute of Chemical Engineers, December 1, 1971, San Francisco, California; "An Assessment of the Role of Volcanic Dust in Determining Modern Changes in the Temperature of the Northern Hemisphere," by Clayton F. Reitan, an unpublished Ph.D. thesis for the University of Wisconsin, Madison, 1971; "Klimatologische Wetterkarten der Nord-hemisphäre fur die Temperatur in Dezennium 1961/1970," by F. R. Pütz, *Berliner Wetterkarte,* 1971, 106:23.

P. 45, photograph courtesy of NASA.

P. 48, photograph by Bruce Roberts, Rapho-Guillumette.

P. 51, photograph by William B. Finch, Editorial Photocolor Archives.

CHAPTER 3

Pp. 54–55, "Charlestown Basin Pollution," photograph by Ellis G. Herwig, STOCK, Boston.

P. 58, from *Glacial and Quarternary Geology* by Richard Foster Flint, copyright © 1971 by John Wiley and Sons, Inc. By permission of John Wiley and Sons, Inc.

P. 60, photograph courtesy of the National Oceanic and Atmospheric Administration.

P. 63, copyright © Norsk Polarinstitut, Oslo (photo S56).

P. 65, photograph courtesy of the National Oceanic and Atmospheric Administration.

P. 67, photograph courtesy of the Atomic Energy Commission.

P. 69, from *Principles of Plant Physiology* by James Bonner and Arthur W. Galston, W. H. Freeman and Company, copyright © 1952.

P. 73, photograph courtesy of Convair, a division of General Dynamics Corporation.

P. 77, from "Smog" by Lowell G. Wayne in *Scientist and Citizen* (now *Environment*), March 1965, Vol. 7, No. 5. Photograph courtesy of County of Los Angeles Air Pollution Control District.

CHAPTER 4

Pp. 84–85, microphotograph showing particulates from ambient air by Harry V. Rhoads, Research Associate, St. Louis University.

P. 86, photograph by U.S. Forest Service.
P. 88, photograph courtesy of S. N. Linzon, Ontario Department of the Environment.
P. 89, from "Acid Rain" by Gene E. Likens, F. Herbert Bormann, and Noye M. Johnson, *Environment,* March 1972. Source: S. Oden, "Nederbordens forsurningett generellt hot mot ekosystemem," I. Mysterud (red.), *Forurensning og biologisk miljovern,* Universitetsforlaget, Oslov, 1971, pp. 63–98.
P. 91, from Clair C. Patterson, with Joseph P. Salvia, "Lead in the Modern Environment," *Environment,* April 1968, 10:72.
P. 93, from "Pigeons: A New Role in Air Pollution" by M. F. Tansy and R. P. Roth in *Journal of the Air Pollution Control Association,* May 1970, 20:308.
P. 95, illustration by Robert Charles Smith, from "Mercury in Man" by Neville Grant, *Environment,* May 1971, Vol. 13, No. 4, 6.
P. 97, adapted from *Data on the U.S. Economy of Relevance to Environmental Problems,* prepared by Barry Commoner, Michael Corr, and Paul J. Stamler for the Committee on Environmental Alterations, American Association for the Advancement of Science.
P. 99, illustration by Robert Charles Smith, from "Earth, Air, Water" by Justin Frost, *Environment,* July–August 1969, Vol. 11, No. 6, 24.
P. 101, adapted from *Data on the U.S. Economy of Relevance to Environmental Problems,* prepared by Barry Commoner, Michael Corr, and Paul J. Stamler for the Committee on Environmental Alterations, American Association for the Advancement of Science.
P. 105, illustration courtesy of the Atomic Energy Commission Health and Safety Laboratory.

CHAPTER 5
Pp. 112–113, personal pollution, photograph courtesy of the American Cancer Society.
P. 116, photograph courtesy of the Office of Educational Services Statewide Air Pollution Research Center, University of California, Riverside.
P. 117, photograph courtesy of the Environmental Protection Agency, Office of Public Affairs, Washington, D.C.
P. 125, photograph courtesy of Bureau of Mines, U.S. Department of the Interior.
P. 131 (top), photograph courtesy of National Smoke Abatement Society, London.
P. 131 (bottom), photograph courtesy of Birmingham *News.*
P. 135, photograph courtesy of Charles E. Reed, M.D., University Hospitals, University of Wisconsin. It first appeared in "Emphysema with Hereditary Alpha-1 Deficiency Masquerading as Asthma" by Sohei Makino, Louis Chosy, Enrique Valdivia, and Charles E. Reed, in *Journal of Allergy,* 1970, 46:40–48.
P. 137, adapted from "Smoking, Air Pollution, and Health" by P. M. Lambert and D. D. Reid, *The Lancet,* April 25, 1970, Figure 3, p. 856.
P. 138, from Raymond L. H. Murphy, Jr. et al., "Effects of Low Concentration of Asbestos," *New England Journal of Medicine,* December 2, 1971, 285:1276.

CHAPTER 6
Pp. 150–151, "Environment Marchers Demonstration Against Automobiles at the Coliseum Auto Show, New York City, November 29, 1972," photograph by Dan Miller, DPI.
P. 153, from Paul W. Spaite and Robert P. Hangebruck, "Pollution from Combustion of Fossil Fuels," *Air Pollution—1970, Hearings before the Subcommittee on Air and Water Pollution of the Committee on Public Works, U.S. Senate,* Part 1, Washington D.C.: U.S. Government Printing Office, 1970, p. 181.

P. 156, *Federal Register,* Vol. 36, No. 67, pp. 6698–99x, and additional information provided by the Office of Air Programs, Environmental Protection Agency, photograph by Paolo Koch, Rapho-Guillumette.
P. 158, from Paul W. Spaite and Robert P. Hangebruck, "Pollution from Combustion of Fossil Fuels," *Air Pollution—1970, Hearings before the Subcommittee on Air and Water Pollution of the Committee on Public Works, U.S. Senate,* Part 1, Washington, D.C.: U.S. Government Printing Office, 1970, p. 174.
P. 164, photograph by Harvey Mudd II.
P. 171, photograph courtesy of the National Center for Air Pollution Control.
P. 175, power plant in Springfield, Missouri, courtesy of Pritchard Products Corporation.
P. 176, adapted from *Data on the U.S. Economy of Relevance to Environmental Problems,* prepared by Barry Commoner, Michael Corr, and Paul J. Stamler for the Committee on Environmental Alterations, American Association for the Advancement of Science.
P. 180, photograph © 1970 by Tim Kantor, Rapho-Guillumette.
P. 183, photograph courtesy of Novosti Press Agency.
P. 189, photograph by K. W. Gullers, Rapho-Guillumette.

Index

Italic numbers refer to illustrations and photographs.

R

S

T

U

A 3
B 4
C 5
D 6
E 7
F 8
G 9
H 0
I 1
J